Science and Spiritual Practices

Rupert Sheldrake

Science and Spiritual Practices

CORONET

First published in Great Britain in 2017 by Coronet
An Imprint of Hodder & Stoughton
An Hachette UK company

This paperback edition published in 2018

3

A CIP catalogue record for this title is available from the British Library

B format ISBN 9781473630093
eBook ISBN 9781473630109

Typeset in Sabon MT by Palimpsest Book Production Limited,
Falkirk, Stirlingshire

Printed in Italy by Elcograf S.p.A.

Hodder & Stoughton policy is to use papers that are natural,
renewable and recyclable products and made from wood grown
in sustainable forests. The logging and manufacturing processes
are expected to conform to the environmental regulations
of the country of origin.

Hodder & Stoughton Ltd
Carmelite House
50 Victoria Embankment
London EC4Y 0DZ

www.hodder.co.uk

In grateful memory of my parents, Reginald and Doris

Contents

Preface

This book is the result of a long journey through the realms of science, history, philosophy, spiritual practice, theology, and religion, as well as physical journeys through Britain and Ireland, Continental Europe, North America, Malaysia, India and other parts of the world. Science and spiritual practices have been part of my life since I was a child, and I have thought about the relationships between them in many different contexts.

I was born and grew up in Newark-on-Trent, Nottinghamshire, a market town in the English Midlands. I had a fairly conventional Christian upbringing. My family were Methodists and I went to an Anglican boarding school for boys.

From a very early age I was interested in plants and animals. I kept many kinds of animals at home. My father was a herbalist, microscopist and pharmacist, and he encouraged my interests. I wanted to be a biologist and I specialised in science at school. Then I went to Cambridge University, where I studied biology and biochemistry.

During my scientific education, I realised that most of my science teachers were atheists and that they regarded atheism as normal. In England at that time, science and atheism went together. An atheist outlook seemed to be part of the scientific worldview, which I accepted.

When I was seventeen, in the gap between leaving school and going to university, I worked as a lab technician in the research laboratories of a pharmaceutical company. I wanted to have research experience. When I took the job, I did not know that I would be working in a vivisection facility. I wanted to be a biologist because I loved animals. But now I was working in a kind of death camp. None of the cats, rabbits, guinea pigs, rats, mice or day-old chicks that were used in the experiments ever left the lab alive. I felt a great tension between my feelings for the animals and the scientific ideal of objectivity, which left no place for personal emotions.

After I expressed some of my doubts, my colleagues reminded me that this was all for the good of humanity; these animals were being sacrificed to save human lives. And they had an undeniable point. All of us benefit from modern medicines, and almost all of these drugs have been tested on animals first. It would be irresponsible and illegal to test untried, potentially toxic chemicals on humans. Humans have rights, so the argument goes. Laboratory animals have almost none. Most people implicitly support this system of animal sacrifice by benefiting from modern medicine.

Meanwhile, I read Sigmund Freud and Karl Marx, who reinforced my atheist views, and when I went to Cambridge as an undergraduate I joined the Cambridge Humanist Association. After going to a few meetings, I began to find them dull, and my curiosity took me elsewhere. The event that has stuck most in my mind was when we were addressed by the biologist Sir Julian Huxley, a leading light of the Secular Humanist movement. He argued that humans should take control of their own evolution and improve the human race by eugenics, namely selective breeding.

He foresaw a new breed of genetically enhanced children, who would be fathered by artificial insemination using donated sperm. He enumerated the qualities that the sperm donors should have in order to create this uplift in humanity: they should be men who came from a long, scientific lineage, who had great personal achievements in science, and who had risen to a position of high esteem in public life. The ideal sperm donor turned out to be Sir Julian himself. I later learned that he practised what he preached.

As an atheist and as a budding mechanistic biologist, I was expected to believe that the universe was essentially mechanical, that there was no ultimate purpose and no God, and that our minds were nothing but the activity of our brains. But I found all this a strain, particularly when I fell in love. I had a beautiful girlfriend, and in a phase of intense emotion, I was going to physiology lectures on hormones. I learned about testosterone, progesterone and oestrogen, and how they affected different parts of male and female bodies. But there was a huge gap between the experience of being in love and learning these chemical formulae.

I also became increasingly aware of the great gulf between my original inspiration – an interest in living plants and animals – and the kind of biology I was being taught. There was almost no connection between my direct experience of animals and plants and the way I was learning about them. In our laboratory classes, we killed the organisms we were studying, dissected them, and then separated their components into smaller and smaller bits, until we got down to the molecular level.

I felt that there was something radically wrong, but I could not identify the problem. Then a friend who was studying literature lent me a book on German philosophy containing an essay on the writings of Johann von Goethe, the poet and botanist.[1] I discovered that Goethe, at the beginning of the nineteenth century, had a vision of a different kind of science – a holistic science that integrated direct experience and understanding. It did not involve breaking everything down into pieces and denying the evidence of one's senses.

The idea that science could be different filled me with hope. I wanted to be a scientist. But I did not want to plunge straight into a career of research, which my teachers assumed I would do. I wanted to take some time out to look at a bigger picture. I was fortunate to be awarded a Frank Knox Fellowship at Harvard, and after graduating from Cambridge, I spent a year there (1963–4) studying philosophy and the history of science.

Thomas Kuhn's book *The Structure of Scientific Revolutions* had recently come out, and it made me realise that the mechanistic theory of nature was what Kuhn called a 'paradigm' – a collectively held model of reality, a belief system. Kuhn showed that periods of revolutionary change in science involved the replacement of old scientific models of reality with new ones. If science had changed radically in the past, then perhaps it could change again in the future – an exciting possibility.

I returned to Cambridge in England to work on plants. I did not want to work on animals, my original intention, because I did not want to spend my life killing them. I did a PhD on how plants make the hormone auxin, which stimulates the growth of stems, the formation of wood and the production of roots. The hormone powder that gardeners use to promote the rooting of cuttings contains a synthetic form of auxin. I then continued with research on plant development as a Fellow of Clare College, Cambridge and also a Research Fellow of the Royal Society, which gave me tremendous freedom, for which I am very grateful.

During this period, I became a member of a group called the Epiphany Philosophers, based in Cambridge.[2] This group was an unlikely confluence of quantum physicists, mystics, Buddhists, Quakers, Anglicans and philosophers, including Richard Braithwaite, who was a professor of philosophy at Cambridge and a leading philosopher of science;[3] his wife Margaret Masterman, Director of the Cambridge Language Research Unit, and a pioneer of Artificial Intelligence; and Dorothy Emmet, professor of philosophy at Manchester University, who had studied with the philosopher Alfred North Whitehead. Four times a year, we lived as a community in a windmill on the Norfolk coast, at Burnham Overy Staithe, for a week at a time. We had discussions about physics, biology, alternative medicine, acupuncture, psychical research, quantum theory, the nature of language and the philosophy of science. No idea was banned.

During this seven-year period, I was free to do whatever research I liked, and wherever I liked. Funded by the Royal Society, I went to Malaysia for a year, because I wanted to study rainforest plants. I was based in the Botany Department of the University of Malaya, near Kuala Lumpur. On the way there, in 1968, I travelled through India and Sri Lanka for several months, and it was a major eye-opener. I found that there were totally different ways of looking at the world for which nothing in my education had prepared me.

When I returned to Cambridge, I continued with my research on plant development, and in particular, I focused on the way the plant hormone auxin is transported from leaves and stems towards root tips, changing the plant as it flows through it. Although this work was very successful, I became more and more convinced that the mechanistic approach was incapable of giving an adequate understanding of the development of form. There had to be top-down organising principles, not solely bottom-up ones.

An architectural analogy for a top-down principle would be the plan of a building as a whole, and a bottom-up explanation would concern itself with the chemistry and physical properties of the bricks, the adhesive properties of the mortar, the stresses in the walls, the currents in the electric wiring, and so on. All these physical and chemical factors are important for understanding the properties of the building, but by themselves they cannot explain its shape, design and function.

For these reasons, I became interested in the idea of biological fields, or morphogenetic fields, or form-shaping fields, a concept first proposed in the 1920s. The shape of a leaf is not only determined by genes inside its cells that enable them to make particular protein molecules, but also by a leaf-shaping field, a kind of invisible plan or mould, or 'attractor' for the leaf. This is different for oak leaves, rose leaves and bamboo leaves, even though they all have the same auxin molecules and the same kind of polar auxin transport system, moving auxin in one direction only, from the shoots towards the root tips, and not in the opposite direction.

When I was thinking about how morphogenetic fields might be inherited, a new idea occurred to me: there might be a kind of memory in nature giving direct connections across time from past to present organisms, providing each species with a kind of collective memory of form and behaviour. I called this hypothetical transfer of memory *morphic resonance*. But I soon realised that this was a highly controversial proposal, and that I would not be able to publish it until I had thought about it much more thoroughly and looked for evidence, a process that might take years.

At the same time, I became increasingly interested in exploring consciousness through psychedelic experiences, which convinced me that minds were vastly greater than anything I had been told about in my scientific education.

In 1971, I learned Transcendental Meditation, because I wanted to be able to explore consciousness without drugs. At the Transcendental Meditation Centre in Cambridge, there was no need to accept any religious beliefs. The instructors presented the process as entirely physiological. That was fine by me; it worked, I was happy doing it, and I did not need to believe in anything beyond my own brain. I was still an atheist and I was pleased to find a spiritual practice that agreed with a scientific worldview, and did not require religion.

I was increasingly intrigued by Hindu philosophy, and by yoga, and in 1974 I had a chance to go and work in India at the International Crops Research Institute for the Semi-Arid Tropics (ICRISAT), near Hyderabad, where I became Principal Plant Physiologist. I did research on chickpeas and pigeonpeas, and was part of a team breeding better varieties with higher yields and greater resistance to drought, pests and diseases.

I loved being in India, and spent some of my spare time visiting temples, ashrams and going to discourses by gurus. I also had a Sufi teacher in Hyderabad, Agha Hassan Hyderi. He gave me a Sufi mantra, a *wazifa*, and for about a year I did a Sufi form of meditation. One of the things I learned from him was that in the Sufi tradition, pleasure is God-given. His religion was not puritanical or ascetic. He wore wonderful brocade robes, was a connoisseur of perfumes, and sat running his fingers through a bowl of jasmine blossoms as he recited poetry in Urdu and Persian. I had always associated religion with a denial of pleasure, but Agha's attitude was completely different.

Then a new thought crossed my mind: what about Christianity? Since my teenage conversion to atheism and Secular Humanism, I had not given it much thought, even though the Epiphany Philosophers were a Christian group, and we chanted psalms together in plainsong every morning and evening at the windmill.

When I asked a Hindu guru for his advice on my spiritual journey, he said, 'All paths lead to God. You come from a Christian family and you should follow a Christian path.' The more I thought about it, the more sense it made. The holy places of Hinduism are in India, or near India, like Mount Kailash. The holy places of Britain are in Britain, and most of them are Christian. My ancestors were Christian for many centuries; they were born, married and died within the Christian tradition, including my parents.

I began to pray the Lord's Prayer, and I started going to church at the Anglican church, St John's, in Secunderabad. I rediscovered Christian faith. After a while, at the age of thirty-four, I was confirmed in the Church of South India, an ecumenical church formed by the coming together of Anglicans and Methodists. I had not been confirmed at school, unlike most of the other boys.

I still felt a huge tension between Hindu wisdom, which I found to be so deep, and the Christian tradition, which by comparison seemed spiritually shallow. Then, through a friend, I discovered a wonderful teacher, Father Bede Griffiths, who lived in a Christian ashram in Tamil Nadu, in the South of India. He was a British Benedictine monk, who had been in India for more than twenty years.

He introduced me to the Christian mystical tradition, about which I knew very little, and to medieval Christian philosophy, particularly the works of St Thomas Aquinas and St Bonaventure. Their insights seemed to me deeper than anything I had heard about in sermons and churches, or in universities. Father Bede also had a profound understanding of Indian philosophy, and gave regular discourses on the *Upanishads*, which contain many of the core ideas in Hindu thought. He showed how the Eastern and Western philosophical and religious traditions could illuminate each other.[4]

While I was working at ICRISAT, I continued to think about morphic resonance, and after more than four years, I was ready to take some time off to write about it. I wanted to stay in India to do this, and Father Bede provided the perfect solution by inviting me to live in his ashram, Shantivanam, on the banks of the River Cauvery, a sacred river.

Father Bede's ashram combined many aspects of Indian culture with Christian tradition. We ate vegetarian food off banana leaves while sitting on the floor; there was yoga every morning, and one-hour periods of meditation in the morning and the evening. I usually meditated in the shade of some trees on the riverbank. The mass in the morning started with the chanting of the Gayatri mantra, a Sanskrit mantra invoking the divine power that shines through the sun. I asked Father Bede, 'How can you chant a Hindu mantra in a Catholic ashram?' He replied, 'Precisely because it's catholic. Catholic means universal. If it excludes anything that is a path to God, it's not catholic, but just a sect.'

I stayed there for a year and a half, from 1978–9, living in a palm-leaf-thatched hut under a banyan tree where I wrote my book, *A New Science of Life*. I then went back to work at ICRISAT on a part-time basis for several more years, spending part of each year in India, part in Britain and part in California.

Back in Britain, I had a wonderful time rediscovering my native traditions. I loved the fact that just as Indians have pilgrimages, Europeans have pilgrimages, too. I went on pilgrimages to cathedrals, churches, and ancient sites like Avebury. It felt like coming home, reconnecting with my native land and with those who had lived there before. I made it my practice to go to church on Sundays, wherever I was, usually in my local parish church. I still do so.

Soon after *A New Science of Life* was published in Britain in 1981, I was back in India working on my field experiments, when I was invited to speak at a conference in Bombay called 'Ancient Wisdom and Modern Science'. I took a few days off from harvesting my crops and went there to give a talk on morphic resonance. While I was at the conference, I met Jill Purce, who was speaking as part of the ancient wisdom programme. Jill had written a book called *The Mystic Spiral: Journey of the Soul,* and she was also the general editor of a series of beautiful books on Art and Imagination, published by Thames & Hudson, which are still in print today.

Jill and I met again a few months later in India, after she had been on a retreat in the Himalayas as part of her practice of *Dzogchen*, a form of Tibetan Buddhism, and later that year we met up again in England, where we came together. We were married in 1985 and have lived in Hampstead, in North London, ever since.

When I met her, Jill had developed a new way of teaching chanting, introducing people to the power of group chant, drawing on traditions from many different cultures and religions. In her workshops she taught, and still teaches, a form of overtone chanting, traditionally practised in Mongolia and Tuva, which makes audible, high, flute-like notes, harmonics of the fundamental tone of the chant. She also shows how chanting can have powerful consciousness-shifting effects and bring people into resonance with each other.⁵

Over the last thirty-five years I have been doing experimental research on plant growth, morphic resonance,⁶ homing pigeons,⁷ dogs that know when their owners are coming home,⁸ the sense of being stared at,⁹ telephone telepathy¹⁰ and a range of other subjects. From 2005–10, I was the Director of the Perrott-Warrick Project for research into unexplained human and animal abilities, funded by Trinity College, Cambridge.

The results of this research have convinced me that our minds extend far beyond our brains, as do the minds of other animals. For example, there seem to be direct telepathic influences from animals to other animals, and from humans to other humans, from humans to animals, and from animals to humans. Telepathic connections usually occur between people and animals who are emotionally bonded.

Such psychic phenomena are normal, not paranormal; they are natural, not supernatural; they are part of the way that minds and social bonds work. They are sometimes called 'paranormal' because they do not fit into a narrow understanding of reality. But the phenomena themselves can be studied scientifically and they have measurable effects. They are about interactions between living organisms, and between living organisms and their environment. However, they are not in themselves spiritual phenomena.

There is a distinction between the psychic and spiritual realms. Phenomena such as telepathy reveal that minds are not confined to brains. But we are also open to connections with a far greater consciousness, a more-than-human spiritual reality, whatever we call it. Spiritual practices help us to explore this question for ourselves.

Jill's work is one of my inspirations for writing this book, because she has developed a way of teaching spiritual practices that includes anyone who is interested, whatever their religion or non-religion. As I found with Transcendental Meditation, and as I have seen over and over again with Jill's workshops, people can learn spiritual practices, and practise them, without having to start by articulating their beliefs or doubts. Their practices can lead to a deeper understanding, but direct experience comes first.

The same principles apply to all the practices I discuss in this book. All of them are open to Christians, Jews, Muslims, Hindus, Buddhists, animists, neo-shamans, people who are spiritual but not religious, New Agers, Secular Humanists, agnostics, and atheists. I myself am a Christian, an Anglican, and I take part in these practices in a Christian context. But all of them are practised by followers of other religions, and also by atheists and agnostics. No religion, or non-religion, has a monopoly of these practices. They are open to everyone.

Many scientific studies have shown that these practices confer benefits on those who do them. For example, people who make a practice of being grateful are, on average, happier than people who do not. I am writing this book because I believe that in our secular age there is a great need to rediscover these practices, whatever one's religion or non-religion.

There are many kinds of spiritual practice. In this book, I discuss a selection of seven, in all of which I participate myself.

Hampstead, London
February 2017

Introduction

All religions have spiritual practices. These practices help to connect people with each other and with forms of consciousness beyond the human level.

Until recently, most atheists and secular humanists took it for granted that these practices were a waste of time, if not dangerously irrational. But attitudes are shifting, especially in relation to health and wellbeing. While medical sciences have made huge advances, they do not confer a sense of meaning or purpose in life, nor are they about improving relationships, or instilling values of gratitude, generosity and forgiveness. We do not expect medicine to do these things. These are all roles that religions play, and they turn out to have major effects on people's health and wellbeing. Recent research studies show that, on average, religious people suffer less from anxiety and depression than non-religious people;[1] they are less prone to suicide,[2] less likely to smoke,[3] and less likely to abuse alcohol or other drugs.[4]

Most of these studies did not disentangle the effects of specific spiritual practices and beliefs, and all religions involve a wide range of practices. Some of these practices can also be carried out in a secular context, including meditation and gratitude. For non-religious people too, these practices turn out to be good for physical and mental health.

In the twentieth century, many people believed that science and reason would soon reign supreme, and religions would wither away. Humanity would ascend to a secular, reason-based social order, liberated from the shackles of ancient dogmas and superstitions.

But rather than dying out, religions have persisted. Islam has not faded away. Hinduism is alive and well. Buddhism's prestige has increased in previously non-Buddhist countries, partly thanks to the Dalai Lama. The practice of Christianity is indeed in decline in most of Europe and North America, but is growing in sub-Saharan Africa, and also in Asia and the Pacific, where there are now more Christians than in Europe.[5] In Russia during the Soviet period, the state was officially atheist and religion was brutally suppressed, but since the communist system ended in 1991, the proportion of Christians in the population has greatly increased. In 1991, sixty-one per cent of Russians described themselves as having no religion, and thirty-one per cent as Russian Orthodox; by 2008, only eighteen per cent said they had no religion and seventy-two per cent said they were Orthodox Christians.[6]

In response to these unexpected trends, there has been a revival of militant atheism. This twenty-first-century anti-religious crusade has been led by the so-called New Atheists, notably Sam Harris, author of *The End of Faith: Religion, Terror, and the Future of Reason*; Richard Dawkins in *The God Delusion*; Daniel Dennett in *Breaking the Spell: Religion as a Natural Phenomenon*; and Christopher Hitchens in *God Is Not Great: How Religion Poisons Everything*.

The New Atheists do not believe in God, but they have a strong belief in the philosophy of materialism. Materialists believe that the entire universe is unconscious, made up of mindless matter, governed by impersonal mathematical laws. Nature has no design or purpose. Evolution is a result of the interplay of blind chance and physical necessity. Consciousness is confined to the insides of heads and only exists inside brains. God, angels and spirits are ideas in human minds: hence they are inside human brains. They have no independent existence 'out there'.

From within this materialist belief system, religion seems like a morass of superstition and irrationality; it represents an evolutionary stage that humanity has outgrown. People who are still religious are feeble-minded or deluded; they should be liberated from the prison of falsehood in which they are trapped, or at least their children should be educated out of it.

The materialist worldview has played a major role in the secularisation of Europe and North America, which has been accompanied by a decline in traditional religious observance, especially among people of Christian backgrounds.[7] In Europe today, only a small minority practises the Christian faith on a regular basis. In Britain, the percentage of regular churchgoers in 2015 was five per cent of the population, down from twelve per cent in 1980.[8] A far higher proportion of the population, forty-nine per cent, defined themselves as having no religion – the so-called 'nones'. In the white population, nones were a majority.[9]

Except in Russia, a decline in Christian faith and practice has occurred almost everywhere in Europe, in both Roman Catholic and Protestant countries. In 2011 in France, historically Catholic, only about five per cent of the population attended a church service weekly,[10] about the same percentage as in Sweden, historically Protestant.[11] Even in countries where the Catholic Church used to be very strong, there have been dramatic drops in religious observance. In the Irish Republic, in 2011 only about eighteen per cent attended weekly mass, down from nearly ninety per cent in 1984.[12] Even in Poland, the most religious country in Europe, weekly church attendance declined to less than forty per cent by 2011.[13]

Most European countries are now predominantly secular and are often described as post-Christian. But the United States is more religious. In 2014, eighty-nine per cent of Americans said they believed in God, seventy-seven per cent identified with a religious faith, and thirty-six per cent attended services weekly. The proportion of atheists was three per cent, much lower than in most of Europe.[14] But even in the US, religious affiliation and observance are declining.[15]

Now everything is in flux. The fundamental assumptions of materialism turn out to be very questionable when examined in the light of advances in the sciences themselves, as I show in my book *The Science Delusion* (called *Science Set Free* in the USA). Meanwhile, the very existence of human consciousness has become increasingly problematic for materialists, who start from the assumption that everything is made of unconscious matter, including human brains. If so, how does consciousness emerge in brains, when it is absent from the rest of nature? This is called the 'hard problem' in the philosophy of mind.

Spirituality outside religion

These declines in religious affiliation and observance do not mean that most people have become atheists. In a survey in Britain in 2013, only thirteen per cent of adults said they agreed with the statement, 'Humans are purely material beings with no spiritual element.' Over three-quarters of all adults said they believed that 'there are things in life that we simply cannot explain through science or any other means.' Even among people who described themselves as non-religious, more than sixty per cent said there are things that cannot be explained, and over a third believed in the existence of spiritual beings.[16]

Whatever people's avowed beliefs, recent studies have shown that spiritual *experiences* are surprisingly common, even among those who describe themselves as non-religious.[17] These include near-death experiences, spontaneous mystical experiences, and revelations while taking psychedelic drugs. The Religious Experience Research Unit in Oxford, set up by the biologist Sir Alister Hardy, asked British people, 'Have you ever experienced a presence or power, whether you call it God or not, which is different from your everyday self?' In 1978, thirty-six per cent said yes; in 1987, forty-eight per cent; and in 2000, over seventy-five per cent of respondents said they were 'aware of a spiritual dimension to their experience.' In 1962, the Gallup organisation asked Americans if they had ever had 'a religious or mystical experience', and twenty-two per cent said yes; in 1994, thirty-three per cent; and in 2009, forty-nine per cent.[18]

These surveys do not necessarily mean that spiritual and mystical experiences are more common than they were; they may reflect a weakening of the taboo against talking about such experiences. Many people used to be afraid that if they admitted to mystical experiences they would be classified as mentally unbalanced. But mainstream psychiatry and psychology are now more open to 'anomalous experiences', and it is culturally more acceptable to discuss them.[19]

Secularism has not led to an extinction of interest in spiritual realms, nor to an eclipse of spiritual experiences.[20] But many people's spiritual interests and experiences now take place outside traditional religious frameworks. For instance, millions practise yoga and meditation in a secular context. New forms of spirituality are emerging that are based primarily on personal experience. They fill a need that atheism cannot satisfy.

The crisis of faithlessness

Hardline atheists such as Daniel Dennett and Richard Dawkins are suspicious of spiritual experiences and tend to dismiss them as delusions of the brain or chemical side effects. But a growing number of atheists and secular humanists are willing to talk about such experiences, and indeed regard them as essential for human flourishing.

The children's writer Philip Pullman, a prominent public atheist, had a mystical experience as a young man that left him with the conviction that the universe is 'alive, conscious and full of purpose'. In a recent interview he said, 'Everything I've written, even the lightest and simplest things, has been an attempt to bear witness to the truth of that statement.'[21]

The philosopher Alain de Botton, who was brought up as an atheist, has come to the conclusion that by abandoning religion, atheists impoverish their lives. In his bestselling book *Religion For Atheists: A Non-Believer's Guide to the Uses of Religion*, he shows how religion satisfies social and personal needs that a purely secular lifestyle cannot.

De Botton was the son of two secular Jews who, he says, 'placed religious belief somewhere on a par with an attachment to Santa Claus . . . If any members of their social circle were discovered to harbour clandestine religious sentiments, my parents would start to regard them with the sort of pity more commonly reserved for those diagnosed with a degenerative disease and could from then on never be persuaded to take them seriously again.'[22]

In his mid-twenties, de Botton underwent what he calls a 'crisis of faithlessness'. Although he remained a committed atheist, he was liberated by the thought that it might be possible to engage with religion without subscribing to religious beliefs. He came to the conclusion that his continuing resistance to religious ideas 'was no justification for giving up on the music, buildings, prayers, rituals, feasts, shrines, pilgrimages, communal meals and illuminated manuscripts of the faiths':

Secular society has been unfairly impoverished by the loss of an array of practices and themes which atheists typically find it impossible to live with . . . We have grown frightened of the word morality. We bridle at the thought of hearing a sermon. We flee from the idea that art should be uplifting or have an ethical mission. We don't go on pilgrimages. We can't build temples. We have no mechanisms for expressing gratitude. The notion of reading a self-help book has become absurd to the high-minded. We resist mental exercises. Strangers rarely sing together.[23]

De Botton says he wants to enrich the lives of atheists by 'stealing' these practices from religion. He turns to religion for insights into how to build a sense of community, make relationships last, overcome feelings of envy and inadequacy, and get more out of art, architecture and music.

Another atheist, Sam Harris, best known for his anti-religious polemics, is at the same time a committed meditator. He spent two years in India sitting at the feet of gurus, and has been initiated into the Tibetan *Dzogchen* meditative tradition. In his book *Waking Up: Searching for Spirituality Without Religion*, he writes:

> Spirituality remains the great hole in secularism, humanism, rationalism, atheism, and all the other defensive postures that reasonable men and women strike in the presence of unreasonable faith. People on both sides of this divide imagine that visionary experience has no place within the context of science – apart from the corridors of a mental hospital. Until we can talk about spirituality in rational terms – acknowledging the validity of self-transcendence – our world will remain shattered by dogmatism.[24]

Harris now teaches meditation in online courses.[25]

Meanwhile, a new atheist church called the Sunday Assembly has been spreading rapidly. It was founded in London in 2013 by two comedians, Sanderson Jones and Pippa Evans. Its services include singing, small-group bonding and uplifting stories. Its motto is 'Live better, help often, wonder more'.[26] Jones describes himself as a 'humanist mystic' and hopes the Sunday Assembly, unlike earlier humanist groups, will develop an ecstatic or charismatic brand of humanism.[27]

Many old-school atheists are willing to admit the validity of feelings of awe and wonder at the universe as revealed by science. But this is almost their only concession to the subjectivity of spirituality. A new generation of atheists and secular humanists are exploring the traditional territory of religion, and trying to incorporate a range of spiritual practices into a secular lifestyle. Meanwhile, the effects of spiritual practices themselves are now being investigated scientifically as never before.

Scientific studies of spiritual practice

In the late twentieth century, from small beginnings in the 1970s, scientists began to investigate a wide range of spiritual practices, including, but not limited to, meditation, prayer, community singing, and the practice of gratitude. In the year 2001, a comprehensive review in the *Handbook of Religion and Health* brought together the findings of more than 1,200 research studies.[28] In this century, there has been a great increase in the amount of research, and a second edition of the *Handbook* published in 2012 reviewed more than 2,100 original, quantitative, data-based studies published since the year 2000. Many more have been published since. The results generally show that religious and spiritual practices confer benefits that include better physical and mental health, less proneness to depression, and greater longevity.[29]

The old-fashioned opposition between science and religion is a false dichotomy. Open-minded scientific studies enhance our understanding of spiritual and religious practices.

In this book, I discuss seven kinds of practice and review scientific studies on their effects. I am not including all possible spiritual practices, but only a limited selection. I intend to explore several more in a subsequent book.

These practices are compatible both with a secular way of life and also with a religious way of life.

The practices themselves are about experience, not about belief. Nevertheless, as I show in the discussions in this book, beliefs affect the *interpretation* of the practices. For example, over many centuries people have meditated within the Hindu, Buddhist, Jewish, Christian, Muslim, Sikh and other religious traditions. They have done so in the belief that their practices are connecting them to a level of more-than-human consciousness.

Materialists deny as a matter of principle the existence of consciousness beyond the human level. They think of experiences while meditating as nothing but changes within brains, confined to the insides of heads. Nevertheless, whatever their belief system, people who practise meditation often receive benefits that enrich their lives.

The seven practices I discuss in this book are common to all religions. All religions encourage gratitude. There is pilgrimage in all traditions – Hindus go to temples dedicated to gods and goddesses, to holy mountains like Mount Kailash, and holy rivers like the Ganges. Muslims go on pilgrimage to Mecca. Jews, Christians and Muslims go on pilgrimage to Jerusalem. In Western Europe, Christians go to Santiago de Compostela, Rome, Canterbury, and Chartres; Irish Catholics go to Croagh Patrick, the Irish holy mountain, and Lough Derg, the sacred lake.

Reconnecting with the more-than-human-world is part of all religious traditions, and all connect in spiritually meaningful ways with plants. Rituals are an expression of spirituality and are found in all religions and secular societies. All spiritual traditions involve chanting and singing.

At the end of each chapter I suggest two ways in which you can gain direct experiences of these practices for yourself.

I

Meditation and the Nature of Minds

Of all the spiritual practices discussed in this book, meditation is the most inward. When meditating, people withdraw from normal activities, and usually sit still with their eyes closed.

In this chapter, I start by discussing what meditation involves. Then, after a brief history of meditation, I discuss research on its effects on physical and mental health, and on how it affects the physiology of meditators and the activity of their brains. I then look in more detail at the experience of meditation and its implications for our understanding of consciousness, both human and more-than-human.

To an outside observer, someone sitting quietly with closed eyes could be praying rather than meditating, and indeed one kind of prayer, contemplative prayer, is a form of meditation. But the internal experience is very different. Most kinds of prayer engage the mind in outward-directed attention, as in praying for other people, and making requests. These kinds of prayer are *about* something. They express intentions. Meditation is not about intentions or requests: it is to do with letting go of thoughts.

I both meditate and pray, and I think of the difference between them as being like breathing in and breathing out. Meditation is like breathing in, directing the mind inwards; and prayer like breathing out, directing the mind outwards. Meditation involves a detachment from normal everyday concerns, with inward-directed consciousness; petitionary and intercessory prayer link the life of the spirit to what is happening in the outer world: such prayer is outward-directed.

There are many techniques of meditation, and various forms are found in all the main religious traditions. Most are practised while sitting, but some involve moving rather than sitting still, as in Zen walking meditation and Qigong, a series of slow, flowing movements combined with deep rhythmic breathing.

The most widely used practices in the modern Western world are derived from the Hindu and Buddhist traditions, and usually involve silently repeating a mantra, a word or phrase, or paying attention to breathing. What happens is that one part of the mind is involved in repeating the mantra or attending to the breathing, while other parts of the mind continue their normal activities. A continuous flow of thoughts and sensations usually engages and preoccupies us. But having an alternative focus of attention on the mantra or the breath interrupts this flow by providing another reference point for the mind.

In this process, meditators notice that thoughts and sensations flood into their minds, one after another, and that they engage with these thoughts and forget all about the mantra or the breathing – until they come back to it again. Then the process begins afresh.

The practice of repeating the mantra or observing the breath relativises and helps detach the meditator from this continual mental activity that otherwise fills the mind. Through practice, it is possible to watch thoughts come and go like clouds passing through the sky, or fish swimming through the water.

Why is this helpful? What's the point? For people who lead busy, action-oriented lives, meditation can seem like a waste of time. It is the opposite of our usual Western tendency to follow the slogan, 'Don't just sit there: Do something!' It is more like, 'Don't just do something: Sit there!'

One of the effects of meditation is an increase in self-knowledge, a greater awareness of the workings of our minds. We might assume that we are fully in charge of our thoughts and our attention. But even a slight acquaintance with the practice of meditation makes us aware how many thoughts insert themselves into our minds and how little control we have over this process. Even people who have practised meditation for many years do not slip instantly into a bliss-filled state of mental stillness. Their minds continue to generate thoughts and images, and their bodies and senses continue to generate sensations, even if they can avoid feeding them with attention and energy.

Meditation is a spiritual practice because it is about living in the present, which can also be experienced as living in the presence of a mind or consciousness or awareness greater than one's own. By contrast, the thoughts that continually flow into our minds take us out of the present, into memories, or desires and fantasies, or resentments about past wrongs, or about intentions for future activities, or about worries about what we ought to have done or ought to do next, or fears about what might happen in the future. All these kinds of thoughts take our minds away from here and now. The practice of the mantra, or the awareness of our breathing, bring us back to the present.

Meditative practices can lead to an enhanced state of consciousness that is experienced as ineffable, too powerful or beautiful to be described. Attempts to translate this experience into cultural and religious frameworks have led to many different terms, including Buddha consciousness, cosmic consciousness, God consciousness, Christ consciousness, true-Self, Formless Void, and undifferentiated Beingness.[1]

Although techniques of meditation grew up within Hindu, Buddhist, Jain, Christian, Jewish, Islamic, Sikh and other religious traditions, meditation can also be practised in a secular spirit, without any religious framework, and in the modern Western world it is commonly used in this non-religious form, either through various derivatives of Hindu meditation, such as Transcendental Meditation, or of Buddhist meditation, as in mindfulness meditation. These techniques are now taught widely in schools, to business people, to members of the US and other armed forces, to prisoners, and to politicians. Dozens of British Members of Parliament have learned mindfulness techniques, and meet weekly to meditate together.[2] Because of its therapeutic benefits, mindfulness meditation is now recommended within the British National Health Service for people suffering from mild or moderate depression, because it has been found to be as effective, and cheaper, than long courses of antidepressant drugs.[3]

A brief history of meditation

The word *meditation* comes from the same Indo-European root as medicine, measure and meter. The basic meaning of its Latin ancestor is 'to attend to', with the related meanings 'to reflect upon', or 'to apply oneself to'.[4]

This modern usage arose only in the nineteenth century through translation of Eastern spiritual writings. In traditional Catholic Christianity, meditation mainly referred to a meditative reading of the scriptures; the closest equivalent to the modern meaning of meditation was called 'contemplative prayer', a form of silent prayer that went beyond thoughts and images.

No one knows when meditation practices first began. Some people speculate that they started among hunter-gatherers sitting around fires and gazing into the flames and the embers. If so, they could be very ancient indeed, since humans began to use fire at least a million years ago.[5] The first actual evidence for meditative practices dates back to about 1500 BC, with an image of a figure seated in the lotus position on a seal found in India.[6] It seems reasonable to assume, as many Indians do, that proto-yogis were meditating in the Himalayas and elsewhere for several thousand years before texts referring to meditation, such as the *Upanishads*, were written down, starting around 800 BC.

The Buddha was born and lived in India and spent years in ascetic and meditative practices with yogis before he finally achieved enlightenment when sitting under a *bodhi* tree. Buddhism became a mass movement in India from the fifth century BC onwards, and in numerous monasteries the monks spent part of their time meditating. Meditation may have evolved independently in China and other parts of Asia, but was much influenced by the spread of Buddhism through the establishment of monasteries. In China, Japan and Korea, meditative practices were established long before the Christian era, and after Tibet was converted to Buddhism in the eighth century AD, meditational techniques evolved in several new ways in remote, high-altitude caves and monasteries. These techniques included spending long periods in complete darkness and isolation, the practice of elaborate visualisations, and dream yoga, involving the cultivation of lucid dreaming, a kind of dreaming in which the dreamers become aware that they are dreaming, as if waking up within the dream state.

Some Jewish scholars think that meditative practices of some kind were well established very early in Jewish history, even at the time of the Patriarchs, and a verse in the book of Genesis about Isaac, the son of Abraham, could refer to a meditative practice. In the King James Bible, the translation of Genesis 24:63 reads: 'And Isaac went out into the field to meditate at eventide.' Meditation has also been practised within the Jewish mystical tradition, Kabbalah, for over a thousand years.

With the growth of Christian monasticism, starting with the monks in the Egyptian desert in the third century AD, most notably St Anthony of the Desert, a range of meditative practices became part of Christian monasticism. In the Eastern Orthodox Churches these methods were widely diffused, especially in the form of the 'prayer of the heart' or the 'Jesus prayer', a very short prayer that calls on the name of Jesus. The repetition of these prayers is very similar to the repetition of mantras in the Hindu and Buddhist traditions. The repetitive use of mantra-like prayers is a common practice in the Roman Catholic tradition, especially using rosaries.

In the Islamic world, Sufi groups encouraged meditation, especially using one of the names of God, repeated over and over, again, like a mantra. This practice is called *zhikr* or *dhikr*.

Some Westerners learned meditative practices in India and in Buddhist countries in the nineteenth century, and were taught within esoteric movements like the Theosophical Society. Meditation spread more widely in the West in the twentieth century through a series of Asian teachers, notably the Indian yogi Paramahansa Yogananda (1893–1952), and the Japanese teacher D. T. Suzuki (1870–1966), who aroused much interest in Zen Buddhist meditation, especially after he settled in New York in the 1950s.[7]

A new era of interest in meditation began in the 1960s as a result of the psychedelic revolution, the rise of the counter-culture and the hippie movement. After the Beatles met the Maharishi Mahesh Yogi (1918–2008) in 1967, organisations teaching meditation became increasingly popular and successful, not least the Maharishi's own Transcendental Meditation movement. In the early 1990s, one of the Maharishi's personal physicians and close associates was the Indian-American doctor, Deepak Chopra.[8] After his break with the Maharishi in 1993, Chopra continued to spread the message of meditation to millions of people in the West. In addition, the Chinese invasion of Tibet in 1950 drove many Tibetan monks and teachers into exile, including the Dalai Lama, with the result that Tibetan Buddhist teachings have been widely dispersed.

Many different forms of meditation are now taught in Western countries, including a range of Hindu-derived techniques, many Buddhist methods, including Tibetan, Zen, and Theravada Buddhist meditation, including *vipassana*, which in its modern form originated in Burma. Vipassana means 'insight into the true nature of reality', and involves being mindful of breathing, feelings, thoughts and actions. Secularised meditation techniques are now very widely taught and are used therapeutically in healthcare systems. Meanwhile several forms of Christian, Jewish and Muslim meditation have been revived and popularised.[9]

One of the pioneers of scientific research on meditation was a cardiologist at Harvard Medical School, Herbert Benson, who began studying Transcendental Meditation in the late 1960s, and summarised his results in his influential book *The Relaxation Response*.[10] Another pioneering researcher, Jon Kabat-Zinn, at the University of Massachusetts Medical School, combined vipassana and Zen practices with yoga to form a training regime called mindfulness-based stress reduction. In the US, there are now hundreds of stress-reduction clinics in hospitals and health centres based on these procedures, to which doctors can refer their patients.[11]

There are also many teachers of mindfulness meditation, and countless newspaper and magazine articles advocate meditation as a stress-reduction and life-enhancement technique. Popular books tell how meditation has changed the authors' lives and the lives of the people they know, and these books encourage their readers to change their own lives by meditating. One of the most engaging is *A Mindfulness Guide for the Frazzled* by the comedian Ruby Wax. Several prominent atheists have also become advocates of meditation, including Susan Blackmore, in her book *Zen and the Art of Consciousness*.[12] Sam Harris, a New Atheist best known for his anti-religious polemics, now teaches meditation in online courses. His book *Waking Up: Searching for Spirituality Without Religion*[13] is intended as 'a guide to meditation as a rational spiritual practice informed by neuroscience and psychology'.

Enormous numbers of people now meditate. In one of the largest and most comprehensive surveys, the US National Institutes of Health found that in 2012, approximately eighteen million adults – eight per cent of the adult population of the US – and one million children were practising meditation.[14]

I first began to meditate in 1971, after learning how to do so from a Transcendental Meditation teacher in Cambridge. I was an atheist at the time, and I liked the fact that this practice did not involve any overt religious beliefs. I could think of it as purely physiological and psychological. However, when I moved to India in 1974 to work in the International Crops Research Institute for the Semi-Arid Tropics in Hyderabad, I realised that meditation practice was part of a much wider religious and philosophical context, and I became increasingly interested in Indian philosophy about the nature of consciousness. When I was living in Hyderabad, I also came to know a Sufi teacher, and started meditating with a mantra-like *wazifah*. Later, I lived in a Christian ashram in Tamil Nadu from 1978–9, where I adopted a Christian form of meditation with the Jesus prayer, and generally meditated for an hour in the morning and an hour in the evening, sitting on the bank of a sacred river, the Cauvery. I also learned the vipassana technique.

I gave up meditating when I had young children. Sitting quietly in the morning was impossible with lively boys around. But I started meditating again when they had grown up. I use a Christian mantra, and usually meditate for twenty minutes in the morning.

Like millions of other people, I find that meditation has a calming effect, helps me to think more clearly and makes me more aware of my mind's workings. From time to time, unpredictably, I have moments of great peace and joy.

The relaxation response and stress reduction

Meditation practices have attracted much scientific attention, precisely because they have been secularised. This would not have happened, or at least not so soon, if they had been seen as primarily religious.

The pioneering research by Herbert Benson and his colleagues at Harvard Medical School in the 1970s focused primarily on the 'relaxation response'. Benson interpreted this response as a reduction in the fight-or-flight reaction to danger, which is associated with the activation of the sympathetic nervous system.[15] Despite its name, the sympathetic nervous system is not about sympathy, but is one part of the unconscious or autonomic nervous system. The other part is the parasympathetic nervous system, sometimes called the rest-and-digest or feed-and-breed system. The two sides of the autonomic nervous system are complementary. The sympathetic system is activated when there is something to be afraid of; the parasympathetic system when there is nothing to be afraid of. At the most basic level, the parasympathetic system has to be dominant if we are to eat, cry, have sex, urinate or defecate.

In acute stress, the fight-or-flight response is triggered by the release of adrenaline, which causes an increase in heart rate and blood pressure, and a decrease in blood flow to the extremities, including the sexual organs, and a slowing of digestion. The fight-or-flight response also increases the level of the hormone cortisol, which reduces the activity of the immune system. (In medicine, cortisol is called hydrocortisone, and it plays a useful role in the temporary reduction of inflammation.) When the danger has passed, these systems return to normal in the relaxation response. However, in chronic stress this physiological arousal persists, which can lead to a decrease in the activity of the immune system and continual anxiety.

Benson's group investigated a range of techniques for inducing the relaxation response, including meditation, breathing exercises, yoga, and muscle relaxation. They also tested the effects of hypnosis, which could elicit the relaxation response when the hypnotist suggested that the subject entered a state of deep relaxation. All these methods led to decreases in oxygen consumption, respiration rate, and heart rate. In patients with high blood pressure, the blood pressure went down.

In some ways, the relaxation response resembled sleep, but whereas in sleep oxygen consumption declined gradually over several hours until it was about eight per cent less than during wakefulness, in meditation it declined by ten to twenty per cent within a few minutes. There was also a decrease in the level of lactic acid in the blood, which fell by around forty per cent within ten minutes of starting to meditate. Lactic acid is normally produced as a result of muscular activity, and in people prone to anxiety, it increases the likelihood of anxiety attacks.

As well as these physiological changes, the relaxation response induced altered states of consciousness that people described as 'feeling at ease with the world', 'peace of mind' and a 'sense of wellbeing' like that experienced after a period of exercise, but without the fatigue. Most people described these feelings as pleasurable.[16] More recent research, discussed below, has revealed how meditation and the relaxation response affect the activity of various regions of the brain, including a deactivation of the 'default mode network', which is associated with rumination and being lost in thought.

Benson's methods were extended very widely through health centres, clinics, and church groups, reaching millions of people from the 1970s onwards. In addition, his book *The Relaxation Response*, first published in 1975, was a bestseller and sold millions of copies.

Benson recommended the use of a focus word or mantra, and advised people to use a word, phrase or short prayer that was personally meaningful, depending on their own religious tradition. He encouraged secular or non-religious people to focus on 'words, phrases or sounds that were compelling to them, such as the words *love*, *peace*, or *calm*'.

Here are Benson's instructions:

1. Pick a focus word, short phrase, or prayer that is firmly rooted in your belief system.
2. Sit quietly in a comfortable position.

3. Close your eyes.
4. Relax your muscles, progressing from your feet to your calves, thighs, abdomen, shoulders, head and neck.
5. Breathe slowly and naturally, and as you do so, say your focus word, sound, phrase, or prayer silently to yourself as you exhale.
6. Assume a passive attitude. Don't worry about how well you are doing. When other thoughts come to mind, simply say to yourself, 'Oh well', and gently return to your repetition.
7. Continue for ten or twenty minutes.
8. Do not stand immediately. Continue sitting quietly for a minute or so, allowing other thoughts to return. Then open your eyes and sit for another minute before rising.
9. Practise the technique once or twice daily. Good times to do so are before breakfast and before dinner.[17]

The other main strand of the modern meditation movement was popularised by Jon Kabat-Zinn, who was introduced to Zen Meditation as a student. He went on to study meditation with other Buddhist teachers, most importantly Thich Nhat Hanh, a Vietnamese Buddhist monk and Seung Sahn, a Korean Zen master. Kabat-Zinn was Jewish by birth, but did not identify with any religion, even Buddhism. Instead, he deliberately secularised the Buddhist teachings he had received. He launched his stress reduction and relaxation program in 1979. This evolved into the practice he called mindfulness-based stress reduction, which has spread worldwide.

The main difference between the work of Benson and Kabat-Zinn is that Benson's recommended technique stems from the Hindu tradition of mantra-based meditation, which is a focused attention method. Kabat-Zinn's mindfulness procedure is sometimes called open-monitoring meditation, because it involves the non-reactive monitoring of experience from moment to moment. Benson's method involves mantras, Kabat-Zinn's does not.

What both kinds of meditation have in common is that they focus attention in the present moment. They create an alternative centre of attention in the form of the mantra or the awareness of bodily sensations, which has a distancing effect on thoughts, feelings, ruminations, fantasies, and worries. These continue to arise, but insofar as meditators return to the mantra, or pay attention to breathing and other bodily sensations, they can come back to present awareness. They are once again in the moment.

Another kind of meditation is derived from the Buddhist technique of *metta*, where the meditator focuses on developing compassion, or a sense of care for living beings. In its secularised form, it is called loving-kindness meditation. Thich Nhat Hanh calls the practice of compassion 'engaged Buddhism' and draws strong parallels with the Christian tradition of loving-kindness. He also links the Christian understanding of the Holy Spirit to the experience of mindfulness: 'We have the capacity to recognise the presence of the Holy Spirit wherever and whenever it manifests. It too is the presence of mindfulness, understanding and love.'[18]

Various forms of meditation are still taught within religious traditions, for example by Tibetan lamas, by Hindu, Jain and Sikh gurus, by Sufi masters, and by Jewish and Christian teachers. Indeed, all forms of meditation are derived from religious traditions.

Health benefits of meditation

Since the 1960s, scientific journals have published thousands of papers on the effects of meditation on health and wellbeing.[19] These effects include a reduction of anxiety, a reduction in allergic skin reactions, less incidence of angina and cardiac arrhythmias, relief from bronchial asthmas and coughs, reduced problems with constipation, fewer problems with duodenal ulcers, less dizziness and fatigue, lower blood pressure in people suffering from hypertension, alleviation of pain, reduction of insomnia, improved fertility, and help in mild to moderate depression.[20]

Studies of school children and college students who meditated showed significant positive effects on social competence and well-being. Even the US Marine Corps tried using 'mindfulness mind-fitness training' to enhance the performance of troops. A journalist who visited a training camp in Virginia described how the rigorous training procedure was interspersed with periods in total silence: 'You'd see men sitting in the lotus position in their field uniforms with rifles across their backs.' The training they received was designed to lower stress, increase mental performance under the duress of war, and improve the capacity for empathy.[21] But this shows how far secular mindfulness meditation has moved from its Buddhist roots. It is hard to imagine how increased military effectiveness can be combined with loving-kindness towards enemies.

Meditation is also helping US military veterans. One study found an impressive reduction in symptoms of post-traumatic stress and depression,[22] and in 2015 meditation was being offered at fifteen Veterans Administration centres.[23]

Many studies have shown that mindfulness meditation is at least as effective, if not more effective, for treating mild to moderate depression than antidepressant drugs.[24] It is also cheaper, and of course has no drug-induced side effects.

This is not to say that meditation is without danger. Out of the many people who try meditating, a small minority have adverse reactions. According to the official National Institutes of Health (NIH) guidelines in the US, 'Meditation is considered to be safe for healthy people. There have been rare reports that meditation could cause or worsen symptoms in people who have certain psychiatric problems.'[25] This is not a new issue. Most spiritual traditions have long recognised that there can be periods of difficulty on the spiritual path, which the sixteenth-century Christian mystic St John of the Cross called 'the dark night of the soul'. This is one reason why religious traditions put a strong emphasis on guidance by a competent teacher. In the secular context, meditation is often portrayed as a self-improvement practice, good for stress reduction and enhanced productivity. In the interests of spreading it quickly, it is often taught through books and online courses, and the personal help of experienced teachers is less readily available.[26]

Nevertheless, for millions of people meditation brings both subjective and objectively measurable benefits, and in the modern Western world the most persuasive measure is money. A recent large-scale study compared thousands of people who received training in a 'relaxation response resiliency program', which included meditation, with thousands of otherwise comparable people who did not have this training, and looked at the medical expenditure they incurred. Over a median period of 4.2 years, those who had received the relaxation training had forty-three per cent fewer medical bills per year, an effect that was highly significant statistically. They also had half as many visits to emergency departments.[27] Meditation can save billions of dollars a year.

Changes in brains induced by meditation

Meditation tends to reduce ruminations, obsessions, cravings, fantasies and being lost in thought. Not surprisingly, these changes in the activity of the mind are associated with changes in the activity of the brain.

During rumination and when we are lost in thought, a linked set of brain regions become active. These are called the default mode network, made up of interacting brain regions that become active by default when someone is not involved in an outward-directed task. This network is involved in daydreaming, mind-wandering, self-centred thinking, remembering the past, planning for the future and also thinking about other people. As meditation proceeds, and as meditators become more experienced, there is a decrease in the activity of the default mode network.

One of the key areas of the brain involved in the control of attention seems to be the posterior cingulate cortex (PCC), which is near the back of the head. Another important part of this network is the medial prefrontal cortex (mPFC) (*Fig. 1*). When attention is focused on a specific task, these areas of the brain are deactivated. When not in a state of focused attention, then they are activated, and link up with the default mode network.

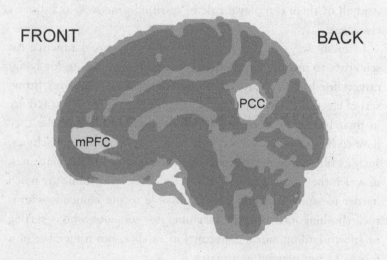

FRONT   BACK

Figure 1. A cross section of the human brain showing the location of the posterior cingulate cortex (PCC) and the medial prefrontal cortex (mPFC).

Meditation is not the quickest way of shutting down the default mode network. Engagement in physically or mentally challenging activities shifts the mind very quickly, focusing attention in the moment. Here is an example, from my friend Gifford Pinchot:

In my forties, I was so intensely involved in my work I could not stop thinking about it, not even after work. Meditation did not change this. However, when rock climbing, after getting above fifty feet up, I thought about nothing but the next moves, if I thought at all. Often it was just a flow of motion in which my body seemed to know what to do next.

Not only rock climbing, but a great many sports or other activities bring people into the present. Engagement in physical work, playing music, looking after young children, singing, dancing and many other activities also switch attention to the present moment. And all of them can play a role in spiritual practices, as I discuss elsewhere.

In many ways meditation is the easiest spiritual practice for scientists to investigate. Meditators are literally sitting (or lying) targets for brain researchers. Either this research involves fitting sets of electrodes to people's skulls to measure the electrical activity in their brains, as in electroencephalographs (EEG), or else they have to lie down and be slid into large, noisy scanning machines, such as functional magnetic resonance imaging (fMRI) machines, in which they must stay very still. People who are mobile are much harder to study. It would not be possible to put someone who is rock climbing into an fMRI machine, nor someone who is surfing or snowboarding, nor a golfer making a shot, nor a member of a football team playing in a match.

Not surprisingly, experienced meditators show bigger changes in their brains when meditating than novices. In a collaborative study with the Dalai Lama, the neuroscientist Richard Davidson arranged for eight experienced Tibetan monks to be tested in his laboratory at the University of Wisconsin. These monks had been trained for an estimated 10,000 to 50,000 hours each, over fifteen to forty years. As a control, ten student volunteers with no previous meditation experience were tested after only one week of training.

The participants were fitted with EEG sensors containing 256 electrodes and asked to meditate for short periods.

Davidson was especially interested in measuring gamma waves, some of the highest-frequency brain impulses detectable by EEGs. Gamma waves range between twenty-five and one hundred cycles per second, and are usually around forty.

The electrodes picked up much greater activation of fast-moving and unusually powerful gamma waves in the monks, and found that the movement of the waves through the brain was far better organised and coordinated than in the students. The meditation novices showed only a slight increase in gamma-wave activity while meditating, but some of the monks produced gamma-wave activity more powerful than any previously reported in healthy people.[28]

The changes in brain activity that occur when people are meditating are not merely temporary; they seem to lead to changes in brain structure as well. In a study at Harvard Medical School by Sara Lazar and her colleagues, the brains of long-term meditators were compared with a control group. The meditators turned out to have more grey matter in the auditory and sensory cortex.[29] As Lazar remarked, this 'makes sense. When you're mindful, you're paying attention to your breathing, to sounds, to the present moment experience, and shutting cognition down.'

There was also more grey matter in the frontal cortex, associated with working memory and decision-making. Lazar said, 'It's well documented that our cortex shrinks as we get older – it's harder to figure things out and remember things. But in this one region of the prefrontal cortex, fifty-year-old meditators had the same amount of grey matter as twenty-five-year-olds.'[30]

Was this simply because of some selection bias in the people that Lazar's team studied? To find out, they recruited participants who had not meditated before, and arranged for them to have eight weeks' training and practice in mindfulness meditation. They then compared the changes in their brains with a control group who did not meditate.

Astonishingly, in just eight weeks there were measurable changes in the brains of the meditators, with increases in the density of grey matter in the PCC (*Fig. 1*), the left hippocampus, the temporoparietal junction (TPJ) at the side of the cortex, towards the back of the brain, and a region of the brainstem called the pons.[31] The researchers speculated that the changes in the hippocampus might be associated with improved regulation of emotional responses, and in the PCC and TPJ with 'the perception of alternative perspectives'. The PCC, as we saw, is involved in the control of attention.

We are not surprised when exercises such as weightlifting lead to physical changes in muscles. Changes in brains as a result of specific mental activities are only surprising because neuroscientists used to believe that brain structures were more or less fixed in adults. But there is now a widespread recognition of neuroplasticity: brains can change.

All these studies on brain activity tell us about brains, but do not tell us what is going on in consciousness. Are the conscious changes that occur during meditation all located inside the head? Or do they involve the connection of the meditator's consciousness with a far greater mind, the source of consciousness itself?

What meditation shows us about minds

In the modern world, all sorts of people meditate, including atheists and agnostics. Probably everyone who meditates agrees about its benefits in reducing stress and in revealing something of the nature of the mind. But opinions differ greatly when it comes to interpreting the moments of calm and joy that go beyond our normal experience.

For materialists, meditation is nothing but an activity of the brain, and therefore all its effects are confined to the brain, including the highest states of mystical experience. At first sight this seems plausible. The practice of meditation does indeed change the physiology and activity of the brain and other parts of the body. It also leads to structural changes in the brain tissue. But this does not prove that the experience is confined to the brain. If I look out of my window at a tree, specific changes occur in my retina, optic nerves and in the visual-processing parts of the brain. But these changes in the brain do not prove that the tree is nothing but a product of brain activity. The tree really exists and it is outside my brain.

The crucial question is whether meditation enables our minds to connect with a mind or consciousness vastly greater than our own. This is what meditators have traditionally believed, and this has been one of the principal motives for meditation. It can help us to transcend our own minds and our own being. The experience of bliss, *nirvana* or *samadhi* is not just about wellbeing, but about experiencing more deeply the nature of reality.

One of the key insights of the *rishis* or seers of ancient India was that our minds are of the same nature as the ultimate consciousness that underlies the universe. For example, in the *Kena Upanishad*, we read:

What cannot be spoken with words, but that whereby words are spoken: know that alone to be Brahman, the Spirit; and not what people here adore.

What cannot be seen with the eye, but that whereby the eye can see: know that alone to be Brahman, the Spirit; and not what people here adore.

What cannot be heard by the ear, but that whereby the ear can hear: know that alone to be Brahman, the Spirit; and not what people here adore.[32]

We know about consciousness by experiencing it. Brahman – or God, or ultimate reality – is not proved by scientific observations of external reality, that which is 'seen by the eye'. Rather our very ability to speak and to see and to hear comes from our participation in this ultimate mind, from which all other minds are also derived. One often-used analogy is buckets of water reflecting the sun. We see a reflection of the sun in each bucket, and each reflection seems separate from all the others, but they are all reflections of the same sun. Likewise, all our minds seem separate, but they are all reflections of the same ultimate mind or consciousness.

Buddhists differ from Hindus in their interpretation of this ultimate reality. Hindus think of it as Brahman, the Lord, or God, or the Spirit, whereas Buddhists avoid calling it God. When the Buddha achieved enlightenment sitting under the *bodhi* tree, he entered a conscious state of *nirvana*, of ultimate peace and blessedness beyond all the changes of this world. But nevertheless, in both traditions this ultimate reality is vastly greater than our brains, and it is not confined to the inside of our heads. Similarly, for Jewish, Christian and Muslim mystics, a direct mystical experience of God is not merely happening inside our brains, but is a direct connection with God's being.

St Thomas Aquinas (1225–74) saw the human experience of blessedness or joy as a sharing in God's being:

> But good in the highest degree is found in God, who is essentially the source of all goodness. And so it follows that the final perfection of human beings and their final good consists of adhering to God . . . Blessedness or happiness is simply the perfect good. Therefore, all sharers in blessedness are necessarily blessed only by sharing in the blessedness of God, who is essential goodness itself.[33]

From this point of view, the benefits of meditation are not simply physiological. Meditation helps bring our minds closer to ultimate reality, which is conscious, loving and joyful. Our minds are derived from God, and share in God's nature. Through meditation we can become aware of our direct connection with this ultimate source of our consciousness, when we are not distracted by thoughts, fantasies, fears and desires. And this contact with the ultimate consciousness is inherently joyful.

Yet the practice of meditation does not necessarily lead to this conclusion.

The ambiguity of secular Buddhism

There is an inherent ambiguity in the modern meditation movement. At one extreme is the use of meditation as a learnable technique to reduce blood pressure, diminish stress, help healing, prevent depression, and give greater psychological insight. Meditation can help people who live frazzled lives. There is plenty of scientific evidence for all this. Mindfulness seems fully compatible with the philosophy of scientific materialism, which locates the mind inside the head. From this point of view, meditation is rather like going to a mental gym for a regular workout.

On the other hand, both the Hindu and Buddhist traditions start from a completely different conception of reality. They see the world as full of suffering, pain and conflict. The only way to get free is through spiritual liberation. Practitioners can escape from the world of suffering through a kind of vertical take-off, leaving the cycles of birth and death behind them. When liberated or enlightened, the consciousness of the seer becomes one with the consciousness that underlies the universe. But in the Mahayana Buddhist tradition, as in Tibet, those who attain this state of liberation, or Buddhahood, are encouraged to become *Bodhisattvas*, voluntarily returning after death to another human life, through rebirth, to help liberate all sentient beings.

Hindus think of this consciousness as the consciousness of God or Brahman. This ultimate reality is *sat-chit-ananda*. *Sat* means being, *chit* consciousness or knowledge, and *ananda* joy. This ultimate consciousness includes the knower – the conscious ground of being – the known, and the joy of knowing and being. Insofar as practitioners experience their minds as absorbed in the being of God, they are joyful because God is joyful.

The Buddhist description of ultimate conscious reality is *nirvana*, enlightenment, or liberation from embodied existence, and absorption into joy and freedom. Meditation is not an end in itself, but part of a path that can lead to liberation.

The Hindu and Buddhist traditions, like other religions, take for granted the existence of realms of consciousness far beyond the human. They see human consciousness as derived from and connected with an ultimate conscious source. By contrast, for materialists, it is all in the brain. There is no such thing as a vast realm of consciousness beyond the human level. This is an illusion, an irrational belief system.

Most secular practitioners of meditation may not notice this conflict. Their focus is primarily on their own lives. But the Secular Buddhist movement makes this ambiguity explicit. Secular Buddhists are practitioners who use Buddhist techniques, but reject Buddhism as a religion. They dissociate themselves from myths about the Buddha's birth, and from beliefs in numerous bodhisattvas, *dakhinis* and other spiritual beings. They reject the idea that nirvana is in any sense 'out there' and exists independently of human minds. They interpret the life of the Buddha as that of a philosopher teaching a way of life, rather than a religious leader.[34]

One of the most extreme exponents of the Secular Buddhist movement is Sam Harris, whom I mentioned earlier. After a secular upbringing and experiences with psychoactive drugs as a student, he dropped out of college and went in search of self-understanding to India, where he studied with a series of gurus for more than two years. He then went back to America, resumed his studies and did a PhD in neuroscience, before launching a new career as a militant atheist. With his first book, *The End of Faith: Religion, Terror, and the Future of Reason*, he achieved international fame as one of the New Atheists. But he is now going further than other anti-religious crusaders. He has found a new way of attacking religion. Instead of denying spirituality, he wants to take it over, and remove it from the realm of religion. In his book *Waking Up: Searching for Spirituality Without Religion*, he writes, 'My goal is to pluck the diamond from the dunghill of esoteric religion.'[35]

Harris's principal teacher was Tulku Urgyen Rinpoche, a Tibetan master who lived for more than twenty years on retreat in a hermitage. His title *tulku* meant that in the eyes of Tibetans, and in his own eyes, he was a reincarnated master, or more precisely the 'emanation' of a deceased master called Nubchen Sangye Yeshe, a ninth-century student of Guru Padmasambhava. Urgyen Rinpoche was a teacher in the Dzogchen tradition, able to transmit the experience of self-transcendence directly to a student. Harris received this transmission from him, and in only a few minutes his life was changed.[36]

However, he denied that there was anything 'supernatural, or even mysterious, about this transmission of wisdom from master to disciple.' Instead, he said, 'Tulku Urgyen's effect on me came purely from the clarity of his teaching . . . I didn't have to accept Tibetan Buddhist beliefs about karma and rebirth or imagine that Tulku Urgyen or the other meditation masters I met possessed magic powers.'[37] But if this astonishing Dzogchen transmission is nothing but a matter of clear teaching, then why cannot Harris, or anyone else, transmit it through books or online courses? In the Tibetan tradition, transmission involves more than words. It needs a living contact. It is a kind of resonance whereby the master is able to induce something of his own conscious state in the person he is initiating.

Harris rejects many of the beliefs of his own teachers as superstitious.[38] He believes that even the Dalai Lama is fundamentally mistaken, because like many other Tibetans he believes in rebirth and consults oracles. Unlike Harris, he has not plucked 'the diamond from the dunghill of esoteric religion'.

Harris's default position is the materialist theory of consciousness advocated by most of his atheist colleagues. Materialists believe that consciousness is nothing but brain activity. However, Harris says that he is not fully committed to this theory. He admits the possibility of consciousness beyond the brain, which all religions accept, but he remains hostile to all religions:

> I remain agnostic on the question of how consciousness is related to the physical world. There are good reasons to believe that it is an emergent property of brain activity, just as the rest of the human mind is. But we know nothing about how such a miracle of emergence might occur. And if consciousness were irreducible – or even separate from the brain in a way that would give comfort to Saint Augustine – my worldview would not be overturned. I know that we do not understand consciousness.[39]

Harris is a sophisticated atheist, and he admits that we do not understand consciousness. But then how can he be so sure that human (and animal) consciousness is all that there is, and that there are no trans-human realms of conscious being? Surely this is no more than an assumption, a belief, an atheist leap of faith.

Most meditators probably just do what they do, and are not motivated to engage in this debate. But this is not solely a theoretical question: it affects people's motivation. Is meditation only about improved health and fitness, increasing a person's ability to get what they want in the world? Is my meditation simply about me? Or is it about linking to a higher, more-than-human realm of consciousness?

The same questions arise in relation to the physical and mental benefits of meditation. Are these purely because of the physiology of the relaxation response and changes in brain activity and brain anatomy? Or, in addition, do some of these benefits flow from connecting with a ground of conscious being beyond individual humans? Religious people acknowledge this connection to a greater consciousness, and its transformative potential. Atheists and secular humanists do not. But as they continue to meditate, their understanding may change, as I found myself.

Mystics in all religious traditions have had direct experiences of being connected to, or absorbed into, more-than-human consciousness. Atheists claim that these experiences are illusions produced inside brains; they assume that they cannot refer to anything beyond the human level. But why not trust rather than reject these direct experiences? After all, the only way we can know about consciousness is through consciousness itself. And we know that one consciousness can link with other consciousnesses, as in our relationships with each other. Through meditation and through mystical experiences, our conscious minds connect with more-than-human conscious minds, and ultimately with the source of all consciousness. Just as we can come into a kind of resonance with each other through love and through taking part in shared activities, so we may come into resonance with more-than-human minds when we are not preoccupied with our own desires, fantasies and fears.

Two meditative practices

MEDITATION

If you already practise meditation, I have no suggestion to add. If you used to practise meditation and have given it up, I suggest starting again, with a daily routine.

If you have never practised meditation, you can try it by

following the procedure for the relaxation response (pp. 32–3) or try methods in one of the many books on meditation. Or you can look for a teacher you respect, and preferably one that is consonant with the rest of your spiritual or religious life. If you are an atheist, you can follow Sam Harris's instructions,[40] or Ruby Wax's,[41] or find one of the many secular teachers. If you are on a religious path, then you will probably feel most comfortable following the instructions of someone in your own tradition. There are meditation teachers in all the major religious traditions, Jewish[42], Christian[43], Muslim[44], and many Hindu and Buddhist teachers. But above all, make a commitment to practise, and try to create a regular routine. If you do not, your practice is likely to be squeezed out by all the demands of your busy life.

If you meditate regularly, then you will start on a journey that can take you far beyond your existing beliefs and limitations, as well as making you happier and healthier.

SPENDING TIME IN SILENCE

The modern world crowds out silence with noise and distractions. We are used to being busy, with our minds racing. In addition, most people are perpetually accompanied by a source of endless distraction in the form of a mobile phone. Meditation is one way of being silent, but there are others. For example, if I am walking in the countryside and talking to someone, I notice much less than if I am walking silently. If I am in a garden, I barely see the plants and hardly hear the birds when I am chatting. If I go to an art gallery, I do not see the pictures as well if I am looking at them while I am talking to other people, because I am bound up in words. Silence helps us to be more open to sights and sounds and smells, and to the world around us.

Hindu yogis and Tibetan sages traditionally meditate in remote caves in the mountains. Jesus used to go to the hills to pray, and Jewish prophets went into the wilderness. Christian hermits and monks often lived in places remote from towns and cities, and some still do. There have always been wild, silent places, and there still are, especially at night. And within towns and cities, many Anglican and Roman Catholic churches are open during the daytime when there are no services going on, and provide oases of quiet. The streets outside may be intense and busy, but there is often a remarkable stillness and peace within these sacred buildings.

Finding times and places to be silent is one of the simplest ways to expand our sensory and spiritual awareness.

2

The Flow of Gratitude

Most of us have been thankful for presents, or for the gifts of love, help and hospitality. We know what gratitude feels like. Everyone is in favour of it, or at least in favour of receiving it. Many children learn at an early age that it is good manners to say, 'Thank you.' Even in cultures where verbal expressions of gratitude are not expected, acts of reciprocity are.

Virtually every language has a word equivalent to 'gratitude' and all major religions encourage expressions of gratitude.[1]

The opposite of gratitude is a sense of entitlement. Our everyday life in a money-based economy heightens ingratitude because there is no need to feel grateful for a service we pay for. If we are staying in an expensive hotel, we feel entitled to a functional plumbing system and clean sheets and towels. We do not feel a need to be grateful for them; we take them for granted. If we pay for a product or a service, it is part of a reciprocal exchange.

When I buy apples from the fruit stall near my home, I pay the asking price in cash. The stallholder and I say 'Thank you' to each other, and we sometimes have a friendly chat. But we both know this is an economic exchange, not a gift. In the automatic check-out system at the nearby supermarket, when I buy food there is no need to say, 'Thank you.' The cash-taking interface is a machine, and the store is part of a corporation whose primary duty is to make profits for its shareholders, who expect regular dividends.

Depersonalisation chokes off gratitude, and consumers soon develop a sense of entitlement; they have a legally enforceable right to expect products or services they have paid for, and to complain when they do not receive what they expect. And they usually feel no gratitude for the land that produces their food, or to the farmers who grow it, or to the people who transport and prepare it, because they are so depersonalised and remote.

Disasters change our perspective. Often parents, or husbands, or wives, or children, or friends are taken for granted. But if they die, especially if they die unexpectedly, their families and friends become aware of how much they depended on them, and how much they received from them. If someone almost loses an eye in an accident, he or she feels very grateful to have eyes, when formerly the eyes were taken for granted. If someone loses a computer or smartphone full of personal information, he or she feels grateful if it is returned, even though it may also have been previously taken for granted. If there is a long power outage, or a strike that prevents goods being delivered to our shops and if we cannot get food supplies, many of us are grateful when our supplies are restored.

As soon as we stop taking almost everything for granted, we begin to realise that we can be grateful for almost everything.[2] We only exist because our ancestors survived and reproduced successfully, right back to the origin of life. As babies, we were totally dependent on other people for our survival. And simply to have survived to the age we are today, we have been supported by hundreds, thousands, even millions of other people: farmers, teachers, builders, electricians, plumbers, doctors, nurses, dentists, grocers, the people who design and make our computers and our smartphones, the pilots and crew who fly us from one part of the world to the other, and so on, and so on.

Then all of us are here only because our planet exists, and life on earth has evolved over billions of years to give us this living planet on which we totally depend.

In turn, our planet is part of the solar system and all life on earth depends on the sustaining light of the sun, and its gravitational pull that keeps us in a steady, life-friendly orbit.

Then the sun depends on the galaxy. It is one cell in the vast body of the Milky Way, along with several hundred billion other stars. At the centre of the Milky Way is a super-massive, hyper-energetic galactic centre, shooting out ionised matter and vast electric and magnetic fields, with magnetic field lines and electric currents in the plasma of the galactic arms millions of light years long, sustaining the environment of our sun.

Our galaxy is part of a galactic cluster, which astronomers call the 'Local Group', consisting of more than fifty galaxies, which is in turn part of the Virgo Supercluster. The electromagnetic radiation permeating the universe includes the light from all the stars and galaxies, some of which we can see with our naked eyes; and coming from all directions invisibly is the fossil light from soon after the Big Bang, known as the cosmic microwave background radiation.

Our scientific creation story tells us that the entire universe originated 13.8 billion years ago in the Big Bang, starting very small, less than the size of the head of a pin; it has been growing and expanding ever since. Some ancient creation myths speak of the origin of all things as the hatching of the cosmic egg, and the contemporary scientific account is similar. Everything comes from a common source, and everything is related. Without this primal creative event, there would be no universe, and we would not exist. And if the properties of subatomic particles, atoms, and the forces of nature had been even slightly different from what they are, there would be no life as we know it, and we would not be here to think about it.

Gratitude and worldviews

Should we feel grateful for all this? The answer depends on our worldview.

If the universe is nothing but an unconscious mechanical system governed by eternally fixed laws of nature, and if evolution occurs through the blind forces of chance and necessity, and if the universe is entirely purposeless, and if biological evolution has no ultimate meaning, then what can we be grateful for or to? The galaxy and solar system were formed automatically and unconsciously through mechanical processes and chance events. Life on earth began through a series of chemical accidents, or perhaps it first appeared on another planet and living germs of life were somehow carried to earth. But however life originated, it has evolved ever since through chance mutations and the forces of natural selection. There is nothing to be grateful for here – and no one to be grateful to. We are lucky, but luck is not a personal force, it is blind chance.

This is the perspective of the believer in contemporary scientific materialism. Most materialists are atheists, and most atheists are materialists. They believe that the entire universe is nothing but unconscious matter, fields and energy, governed by impersonal mathematical laws. Everything happens automatically. All evolution is unconscious.

Meanwhile, as brains grew bigger, minds evolved in advanced animal lines, and most of all in humans. But however wonderful human minds may be, they are nothing but the physical activity of brains, confined to the insides of heads. They are extinguished when brains die. All religious ideas about the conscious survival of bodily death are fantasies.

Through our minds, we can form models of the whole of nature, including a vision of the vastness of the universe, and of its great age. Our theories reach much further in space and time than our unaided senses, but these scientific models are products of human minds, and hence can only exist as conscious thoughts inside human brains. If and when humans become extinct, these theories will become extinct as well, unless humans can pass them on to some other species that survives.

Thus for materialists, although nature is mathematically and physically amazing, it is not deserving of gratitude, because it is not a gift, or an act of choice or purpose, but an inevitable consequence of blind laws and forces. It *should* be taken for granted. So should the existence of minds, imaginations, and scientific thought itself. There is no one to thank for this. To feel gratitude to nature, or the cosmos, or to the power of creativity is to fall prey to anthropocentric thinking, attributing being, purpose or meaning to inanimate nature. This is permissible in romantic poetry, as long as we understand that it is merely a manner of speaking. As far as objective, scientific truth is concerned, we have no need to feel grateful to nature or the source of nature: instead we should feel grateful to the great scientists who have elevated us to our higher and more objective point of view. Gods do not exist, but through science and reason, humans now have godlike powers.

By contrast, in many religious cosmologies, the entire universe has come forth through the creative power of God. In one of the Hindu interpretations, the world is the dream of the god Vishnu, and is all in his mind. In the Judeo-Christian tradition, the primary metaphor for divine creativity is speaking. Words give structure, form, meaning, and interconnection. To be spoken, they require breath. As we know from our daily experience, the flow of the out-breath propels all my spoken words and yours. The word for breath in Greek is *pneuma*. This word means not only breath, but wind. The equivalent Hebrew word *ru'ach*, which is feminine, can be translated as wind, or breath, or spirit. In China, *qi* or *chi* has a similar meaning, and in India *prana*. In science, this universal flow of activity is called energy. God continually creates the world, and creates us, and our minds, through the flow of cosmic energy and through the creation of forms, patterns and meanings.

If we believe that God is the source of all things, and that God's being sustains the universe – a belief shared by Christians, Muslims, Jews, and Hindus – then our ultimate gratitude is to God for the very fact of existence. Buddhists, Taoists and Confucians do not use the word God in the same sense, but all have their own conceptions of ultimate reality. Our gratitude is also due to the universe, our galaxy, our solar system, our earth, on which our lives depend, and the microbes, plants and animals that provide our food, and the human societies and cultures that sustain all human lives.

In religious traditions, there are many ways of giving thanks to the ultimate source of everything, or God.[3] The Jewish psalms, in the Old Testament, are full of praise and thanks to God. In Christian services, these same psalms play an integral part, and there are many specifically Christian hymns of praise and forms of thanksgiving. Traditionally, Christians said grace before meals, and some still do. The same goes for Jews, and for Muslims, and for people in many other religious and national traditions. In the United States, the Thanksgiving festival is an important part of the national culture.

For a materialist or an atheist this is all nonsense, or at best a kind of poetic make-believe. Reality is not a gift of God. Nor are harvests or fruits of the land. They have come about through chance and necessity, and through human science, technology and hard work. Even the care of parents for their children is a genetically programmed response, manipulated by selfish genes that are interested only in propagating themselves. So there is no need to feel grateful even for parental love, since it is programmed by the genes for their own selfish ends.

Personal differences

Research by psychologists has shown what we all know anyway: some people are temperamentally more grateful than others. A familiar way of recognising this difference is through the way different people respond to a glass of water filled to fifty-per-cent capacity. For the grateful, the glass is half full. For the ungrateful, the glass is half empty. Of course, most of us are both grateful and ungrateful, or optimistic and pessimistic, in different situations. And sometimes it is important to be pessimistic. If I notice that the petrol tank in our car is almost empty, it is better to expect the worst, and try to fill it up soon, rather than pressing ahead in the hope that petrol will materialise magically, or that the petrol gauge is wrong.

Since around the year 2000, gratitude has been studied scientifically thanks to the growth of positive psychology. Psychologists have developed questionnaires and scales by which they can rate people's gratitude or ingratitude. They can also assess their well-being and happiness. Study after study has shown that people who are habitually grateful are happier than those who are habitually ungrateful; they are less depressed, more satisfied with their lives, have more self-acceptance and have a greater sense of purpose in life.[4] They are also more generous.[5]

These are correlations. Are happy people grateful because they are happy? Or are people happy because they are grateful?

Positive psychologists have done experiments to try and find out. In one kind of test, the participants were divided at random into three groups. In one group, people were asked briefly to describe five things they were grateful for in the previous week. In another group, people were asked to describe five hassles from the previous week, while in the third group they were asked to describe five events that had affected them in the previous week. These exercises were repeated for ten weeks.

In the gratitude group, the test group, a wide range of experiences led to gratitude, including cherished interactions, good health, overcoming obstacles, and simply being alive. The researchers found that:

[P]articipants in the grateful condition felt better about their life as a whole and were more optimistic about the future than participants in either of the other comparison conditions [those who described hassles or just wrote about events]. In addition, those in the grateful condition reported fewer health complaints and even said that they spent more time exercising than control participants did.[6]

Other experiments that involved counting blessings gave similarly impressive positive results.[7]

In another kind of test, one group of participants was asked to write a letter of gratitude to someone who had helped them in their lives but who they felt they had never properly acknowledged, and then to deliver this letter in person. The control group was asked to write about their early memories. The gratitude group showed a large increase in happiness compared with the control group, lasting for at least a month.[8]

There are now many self-help books on how to be more grateful, how to count your blessings and how to improve your relationships through gratitude. These methods work: not for everyone all the time, but for many people much of the time.

What's wrong with gratitude?

Grateful people are generally happier than ungrateful people, and also tend to be more liked by others. So are there any disadvantages to being grateful?

Perhaps there are. Gratitude, like other human emotions and dispositions, can be exploited by others. In our capitalist society, there has been a big incentive for corporations to learn from positive psychology in general, and from gratefulness research in particular. Many companies give free copies of positive psychology self-help books to their employees; some sponsor training courses and motivational lectures. Having positive, compliant, and grateful employees is good for business. And if employees have to be laid off, then people trained in positive psychology can be hired to try and make them feel that losing their job is a great career opportunity, so that they feel little or no resentment against the company that fired them.[9]

The writer Barbara Ehrenreich is very critical of these practices because she feels they offer solutions in the form of mental disciplines that involve screening out negative thoughts, rather like old-style Calvinists keeping watch for sinful thoughts. Ehrenreich argues that we need negative thoughts if we are to fight against injustice and environmental destruction.[10] I agree with her.

Gratitude can be manipulated and abused, just as love can be manipulated and abused, but this is not an argument against the importance of gratitude in itself. In general, it is better to be loving and grateful than unloving and ungrateful. Most people prefer to be with loving and grateful people than unloving and ungrateful people. But being compulsively loving and grateful can blind us to danger and to destructive human behaviour, and stop us doing something about them. We need an appropriate balance.

Gifts and obligations

We are often resistant to receiving gifts. Why? Because receiving gifts implies an obligation to give something in return. We feel indebted. Lobbyists and marketers know that giving gifts induces a sense of reciprocal obligation that they can use to their advantage.

A seminal book on this subject is called *The Gift*, first published in France in 1953, by Marcel Mauss. He showed that in a wide variety of traditional societies there was no such thing as a free gift. When one tribe or clan gave gifts to another, the receivers were under an obligation to give gifts in return relatively soon. Sometimes they were expected to give them with interest, as occurred in the Native American cultures of the Pacific Northwest. If one group gave ten blankets one year, the other felt obliged to return twenty the next, and so on. This ceremonial gift-giving, or 'potlatch' system, often became competitive and unsustainable, and one outcome was the conspicuous destruction of accumulated wealth, including burning the houses of princes, putting slaves to death, burning precious oils, and casting precious copper objects into the sea.[11]

In religious contexts, sacrificing human or animal lives, or food, or other gifts to goddesses, gods or God was – and still is – a way of reciprocating the gifts received from them. Or maybe it is a way of giving something, expecting a gift in return, as summarised in the Latin phrase *do ut des*, I give in order that you give.

Although our secular modern lives are so strongly influenced by the money-based economy, we are still familiar with the dynamics of gifts and reciprocity. Gifts create social bonds. We are all aware that at least a conventional expression of gratitude, like saying 'Thank you', is part of normal social behaviour.

If gratitude is a social virtue, ingratitude is a social vice. In many social circles, people who are ungrateful are unpopular. A sense of entitlement, or failure to appreciate others properly, is tolerated in aristocrats, oligarchs, and other powerful people in hierarchical societies. But in more egalitarian social groups, a sense of entitlement is not a recipe for being liked. Some people feel entitled to be served and helped by others, without expressing gratitude. In very young babies, this is the only way they can exist. But by the age of about six weeks, most babies begin to smile, and their smiles are often thanks enough. Many young children are soon taught to express their gratitude verbally, and in other ways.

What if our life is a gift of nature? What if nature herself is a gift? Then we have our deepest obligation to the powers that brought us and everything else into being, and our greatest cause for gratitude.

In all religious traditions, hymns of praise and expressions of gratitude towards the divine source of all being are part of reciprocal interaction; thanksgiving links us to the source of the gift of life itself, and all the other blessings in our lives. And one expression of this gratitude is to share our gifts with others, to become part of the flow on which we ourselves depend.

The downside is that by recognising our total dependence on powers beyond ourselves, we can be filled with an overwhelming sense of religious obligation, and guilt at not fulfilling that obligation. One way out of this sense of inadequacy is to become an atheist. If everything happens automatically and unconsciously, if there is no purpose or providence in the world, then there is nothing to feel grateful for.

But this liberation comes at a high price. Being ungrateful is often accompanied by unhappiness. For believers in the materialist theory of nature, living unhappily can seem like an act of heroism, an unflinching fidelity to objective truth. But philosophical materialism is not The Truth: it is a worldview, a belief system. Though it has many committed followers, to believe in it is not a matter of intellectual, logical necessity, but a matter of ideology, or personal or cultural habit.

Grace and gratitude

The Latin word for gratitude is the noun *gratia* from which our word 'grace' is derived. The closely related Latin adjective *gratus* means 'pleasing'. And from these roots come a range of English words including 'graceful', 'disgraceful', 'gracious', 'gratification', 'gratuitous' and 'congratulate'.[12]

Grace itself has several meanings. First, in Christian theology, it is a gift of God, a divine favour. For instance, in the prayer 'Hail Mary, full of grace', Mary is a channel of God's grace, which flows through her and through her womb:

Hail Mary, full of grace, the Lord is with thee.
Blessed art thou among women, and blessed is the fruit of thy
womb Jesus . . .

Second, grace also refers to proportions or actions that are pleasing,
as in graceful movements, or gracious manners. Likewise, it means
attractiveness or charm, as in elegant proportions. In Greek
mythology, the Graces, sister goddesses, were bestowers of beauty,
and were of exquisite beauty themselves.

Third, grace also means thanks, or gratitude, as in saying grace
before meals. There are similar words for thanks in other languages
– in French, *grace à*, 'thanks to'. The word for 'thank you' in
Spanish is *gracias*, and in Italian, *grazie*.

What unifies these meanings is a sense of free flow in both
directions, and from this flow comes graceful movement or beauty.
The giver and the giver of thanks are connected together; they
are in a mutual relationship, as in a bond of loyalty, love and
trust. This is also a theological description of the relationship
between God and those who love, trust and give thanks to him.
Likewise, it describes the widespread pattern of human interaction through the reciprocal giving of gifts and services. This
mutuality of giving underlies relationships in families and other
social groups.

These mutual relationships still exist, and they long preceded
organised trade and money-based economies. The process of
buying and selling quantified and systematised systems of reciprocal giving. But whereas buying and selling follow quantified
rules, giving gifts, gratitude and giving something back are voluntary. They are more free, more personal, and less automatic and
unconscious. Spirit (which I think of as the flow of conscious life)
flows through us when we give, and when we give thanks. In my
view, this flow is a fundamental aspect of all human societies, and
also of human relationships with ancestors, saints, spirits, angels,

gods, goddesses, and ultimate reality, which Jews, Christians and Muslims call God.

The bestselling author and neurologist Oliver Sacks was an atheist, who was put off God as a young man when his Jewish family disapproved of his being gay. His last book, published post-humously in 2015, is called *Gratitude*, and was written when he knew he was dying of cancer. He summed up his feelings as follows:

> I cannot pretend that I am without fear. But my predominant feeling is one of gratitude. I have loved and been loved. I have been given much and I have given something in return . . . Above all, I have been a sentient being on this beautiful planet, and that in itself has been an enormous privilege and adventure.[13]

Being grateful makes us part of this mutual, life-enhancing flow. Being ungrateful separates us from it. When we are part of this flow, we generally feel happier than when we are not part of it, whether we call ourselves atheists or not.[14]

The practice of gratitude connects us with the graceful flow of giving and returning thanks in the human realm and also to the flow of life in non-human nature: in plants and animals, in ecosystems, in the earth, in the solar system, in our galaxy and in the entire cosmos.[15] And if we are open to it, gratitude can connect us directly to the conscious source of all being, of all consciousness, form and energy, which Jews, Christians and Muslims call God, and which Hindus call *Sat-chit-ananda* – Being-Consciousness-Bliss.[16]

Two ways of practising gratitude

COUNT YOUR BLESSINGS

Try to make this a regular practice, for example, every day before you go to bed. Or once a week: on Friday if you are Muslim, on Saturday if you are Jewish, and on Sunday if you are Christian by ancestry. Other traditions have their own special days. If it helps you to write things down and make a list, then do so. You can give thanks for your own life and health, for your family, your teachers and other people who have helped you, for your language, culture, economy, education and society, for plants and animals, to Mother Nature in her many forms, to the entire universe, and to the source of all being. This practice connects you to that which is given to you. The greater your gratitude, the greater the sense of flow, and the greater your desire to give.

SAY GRACE BEFORE MEALS

In my own home, we hold hands around the table before we eat. Sometimes we sing a grace; sometimes someone says a grace, and sometimes we spend a short period in silence together. If I am on my own, I give thanks silently. I suggest making this a practice in your own home.[17] If some of the members of your family or your friends are not comfortable with a grace, then hold hands silently. Or provide a space where anyone can express their thankfulness in their own way, through speaking or singing.

3
Reconnecting with the More-Than-Human World

Human and non-human nature

We are part of nature. We could not exist without the earth, the sun, our galaxy and the whole cosmos. The history of our galaxy reaches back billions of years, and is grounded in the evolution of the universe.

We are also aware of our separation from the rest of nature. There is a distinction between the human world – our social and economic environment, the languages and cultures we inherit, the houses and cities we live in, the computer screens we interact with, the vehicles we travel in – and the rest of nature. Clearly there is vastly more non-human nature than humanised nature in the universe, including billions of galaxies beyond our own.

We are not alone in making a distinction between our own species and the rest of nature. If only for purposes of breeding, members of other animal species have to recognise each other. A peahen has to be able to recognise a peacock, even though he differs in appearance from herself, in order to mate. And social animals, like ants in their nests, or wolves in their packs, recognise and interact intensely with other members of their group. The group itself has a boundary that separates it from the rest of the world, though it interacts continually with its environment. There is an implicit distinction between the group and the wider world on which it depends.

An illuminating way of thinking of non-human nature is in terms of the more-than-human world, a phrase introduced in the

1990s by the cultural ecologist David Abram.[1] It is only because of the more-than-human world of the earth, the solar system, and the entire cosmos, that we are here.

Unfortunately, many of us have acquired a habit of thinking of the rest of the natural world as somehow *less* than human. Galaxies, planets, biological species, molecules, atoms and sub-atomic particles are all mapped out in our scientific theories. We seem to see nature from the outside, as if we are disembodied minds. In schools, children learn about the solar system from books and models, without going out at night to look at the actual planets and constellations. In kindergarten, young children may interact with living animals and plants, but as they grow up, the study of biology becomes increasingly removed from experience. It soon leaves actual animals and plants behind and centres on textbook diagrams of physical and chemical mechanisms, representations of DNA molecules, brain scans and computer simulations.

The scientific models seem more important than the living organisms themselves. And then these models depend on physical processes that can be modelled mathematically, and soon mathematics seems like the ultimate reality. Living nature is replaced by mental abstractions, found only in minds or in computer software. In fact, an understanding of these mathematical models exists only in a tiny minority of human minds, those belonging to mathematicians, mathematically trained scientists and computer programmers.

Yet for many of us, a sense of direct connection with the more-than-human world is of vast importance, and helps to inspire us spiritually.

Everyday connections

Our species, *Homo sapiens*, is thought to be about a hundred thousand years old, and is descended from previous hominin species going back several million years to an ancestor we shared with apes. For the vast majority of hominin history, our ancestors lived in groups and sustained themselves by gathering plants, and sometimes by eating other animals. They were hunter-gatherers.

Hunter-gatherers take it as given that the world around them is alive: the animals and plants, the earth, the heavens, the sun and the moon, the rivers, the sea, the winds, and the weather. They are animists. Their mythologies emphasise the life and inter-connectedness of the natural world, a continual dialogue of souls between humans and non-humans.[2]

Amerindian mythologies presuppose that there was an original spiritual unity, a primal human being, from which all things are derived. Humans do not come from animals; instead, animals come from human-like beings. The South American anthropologist Viveiros de Castro points out that this traditional view is in many ways the opposite of ours. We see human nature as originally animal, and human culture controls our animal nature. Having been animals, we remain animals 'at bottom'. By contrast, 'Amerindian thought holds that, having been human, animals must still be human, albeit in an unapparent way.' The inner nature of all animals is like ours, but their bodies have non-human forms.

The myths of the Campa people of the Peruvian Andes tell how the primal Campa people became irreversibly transformed into the various species of plants and animals. The development of the universe was primarily a process of diversification, and humankind the primal substance from which all things arose. The present-day Campa people are descendants of the ancestral Campa, but are the only ones who escaped being transformed.[3]

By contrast, from the materialist point of view, the entire cosmos is unconscious, with no purpose or meaning. Biological species are genetically programmed machines. We humans have emerged as a result of blind physical processes. Solar systems, planets, animals and plants are mindless mechanisms impelled by physical and chemical forces. Any attempt to see mind, or soul, or psyche, or purpose in non-human nature is nothing but a projection of human minds onto the rest of the world, found in primitive and religious people, and also in children. Secular, modern, scientifically educated, progressive people have grown out of it. Or at least, they *should* have grown out of it.

These different worldviews lead to very different relationships with the rest of nature. If nature is unconscious and mechanistic, as the materialist philosophy assumes, then our scientific understanding is the supreme conscious reality. Our subjective experiences are by-products of the activity of brains.

But according to the evolutionary biologist Edward O. Wilson, himself a secular humanist, it is a mistake to erect a barrier between our subjective experiences and the natural world. Wilson thinks humans have an instinctive need to connect with animals and plants, based on a long evolutionary history as hunter-gatherers. He calls this instinctive love of nature *biophilia*, from the Greek *bios* = life and *philia* = love of. Inherited biophilia underlies 'the connections that human beings subconsciously seek with the rest of life.'[4]

Even in modern industrial civilisations, many people experience a conscious connection with non-human nature. Many have felt that they have been in contact with a greater presence, or mind, or being, or spiritual reality, or God.

For our ancestors – thousands of generations of hunter-gatherers and many generations of farmers – daily connections with animals, plants, landscapes and weather were essential aspects of life. Even though most people now live in cities, keeping pets is a mainstream activity. Pets became widespread only with the onset of large-scale industrialisation and urbanisation in the nineteenth century. Pet keeping can even be seen as a kind of lament for a lost closeness to nature.[5] And even though most people no longer need to grow their own food, many urban people have gardens, allotments, window boxes, houseplants or cut flowers. Among town and city dwellers, millions of people still connect with the non-human world through walking in parks, woods and the countryside, and many millions spend holidays beside the sea. Many people find great satisfaction in working outdoors, and large numbers volunteer to work on organic farms, in woods and on conservation projects.[6]

Benefits of exposure to more-than-human nature

The effects of exposure to the natural world have been studied scientifically.[7] According to a recent summary of this research, 'Nature improves mental health – people are less depressed when they have better access to green spaces. The beneficial effect is not just a matter of physical exercise, although that is part of the picture. There is something about natural environments that improves people's wellbeing . . . Put simply, being in nature feels good.'[8]

Studies in Japan of the practice of 'forest bathing' (*shinrin-yoku*) have shown that wandering in woods had calming physiological and psychological effects, including a reduction in the levels of the stress hormone cortisol in the blood,[9] and an enhancement of activity of the immune system.[10] In summary, 'forest visits promote both physical and mental health by reducing stress.'[11]

In a recent study in Stanford, California, participants, randomly assigned, went on a fifty-minute walk, either in an urban or a natural environment. They were given a series of psychological tests before and after their walk. Those who went on the nature walk were less anxious and with fewer negative ruminations than before the walk, while those who went on urban walks showed no change. The nature walkers' working memory improved, too.[12] In short, they were happier and more attentive.

Hard-core scientists do not trust subjective impressions alone. They like to see what is happening in the brain. In a follow-up study, the Stanford researchers scanned the brains of participants after urban or nature walks. Those who walked in a natural setting had a reduced tendency to brood and, sure enough, the region of the brain most associated with brooding, the subgenual prefrontal cortex, was less active in those who had walked in a leafy, natural setting than those who, walked along busy roads.[13]

These conclusions would not have surprised the founders of the conservation movement in the nineteenth century, or those public-spirited campaigners and visionary town planners who gave our cities their parks and other green spaces. They understood clearly enough the need of areas for outdoor recreation, as discussed below.

However, even though there are many available green spaces, a recent survey in the UK indicated that sixty per cent of the population do not spend any time 'near nature' in a given week.[14] This is despite the fact that the official government policy is 'to strengthen connections between people and nature, and in particular for every child to be able to experience and learn in the natural environment.'[15] As a step towards implementing this policy, the government commissioned a large-scale study in 2013–14 of children's outdoor activities in England. A majority (eighty-eight per cent) 'visited the natural environment' at least once in the previous year; seventy per cent visited at least once a week. Children from high-income households spent more time outdoors than children from low-income households. Not surprisingly, the oldest children in the survey (thirteen- to fifteen-year-olds) made most visits to parks, playgrounds or playing fields with no adults present. About eleven per cent of visits were to local woodland, ten per cent to local rivers or lakes, and seven per cent to the local countryside. However, about twelve per cent of children (about 1.3 million children in England) spent almost no time outdoors.

In the United States, the writer Richard Louv has called this disconnection between children and the natural world 'Nature-

Deficit Disorder'. In his book *Last Child in the Woods: Saving Our Children from Nature-Deficit Disorder*, he linked this lack of connection to childhood trends such as attention deficit disorder, depression and obesity. He found that average eight-year-olds were better able to identify cartoon characters than trees or animals in the neighbourhood. A typical fourth-grade child said, 'I like to play indoors because that's where all the electric outlets are.'[16]

Louv summarised research showing that environment-based education helps children develop better skills in problem-solving and in making decisions. And they have more fun. But unfortunately there are even more incentives for children to stay indoors. By 2016, many American children were spending five to seven hours on screens every day,[17] and children in the UK up to six hours.[18] Contemporary official guidelines recommend that children should not start using screens until they are two years old. But many do. Children, and even toddlers, are in uncharted territory. The artificial world of screens and social media is engulfing them to an unprecedented degree. This is a vast, uncontrolled experiment with the future of humanity.

Children's connections with nature

Urbanisation, the growth of digital media and parental fears mean that most children spend less time outdoors than in all previous generations. But there is no doubt that, given the chance, many feel a sense of connection with the more-than-human world.

The Religious Experience Research Unit at Oxford, founded by the evolutionary zoologist Sir Alister Hardy in the 1960s, collected many thousands of accounts of spiritual experiences. Out of this large sample, about fifteen per cent started with some reference to experiences in childhood.[19] On further questioning, most of the authors of these accounts said that these childhood experiences felt exceptionally authoritative and significant. Most said that at the time they were unable to talk about their experiences to teachers or family members. As one person put it, 'This inner knowledge was exciting and absorbingly interesting, but it remained unsaid, because, even if I could have expressed it, no one would have understood.'[20]

Here is one example of a remembered childhood experience from the Religious Experience Research Unit collection:

[When I was a child] I seemed to have a more direct relationship with flowers, trees and animals, and there are certain particular occasions which I can still remember in which I was overcome by a great joy as I saw the first irises opening or picked daisies in the dew-covered lawn before breakfast. There seemed to be no barrier between the flowers and myself, and this was a source of unutterable delight.[21]

Other respondents spoke of 'feeling a timeless unity with all life', 'a deep and overwhelming sense of gratitude', and 'a sense of unending peace and security that seemed to be part of the beauty of the morning'.[22]

But do these recollections, written in middle or old age, accurately reflect the experiences of childhood, or are they seen retrospectively through rose-tinted glasses? A British teacher, Michael Paffard, tried to answer this question by asking his teenage students to fill in a questionnaire and write an account of their own experiences of joyful or awe-inducing connection with nature, if they felt they had any relevant experiences to describe. Out of four hundred students, fifty-five per cent attempted to describe experiences that could be classified as nature-mystical. Key words used to describe their experiences included 'joyful', 'serene', 'ecstatic', 'holy', 'blissful', 'uplifting', 'timeless' and 'peaceful'.[23] The kinds of experiences they described were very similar to those recalled by much older people.

As well as feeling exhilarated, some of the respondents also felt afraid when confronted with the sky, or the mountains, or the sea, or uninhabited spaces. A sixteen-year-old boy who went to a boarding school wrote:

> I live in Essex on the edge of a vast expanse of salt marsh . . .
> Often in the autumn months I go and sit on the sea wall and
> spend the evening looking across the marshland. When I am away
> at School I long for its lonely wildness and the sense of freedom
> and the strength of nature that it gives. Yet it is terrifyingly deso-
> late when it gets darker and the sea begins to rise and I must get
> away. But I always have to return to it.[24]

In Paffard's study, in sixty-seven per cent of the accounts, the young people were alone when they had these experiences, and most of them took place in the evening or at night.[25] Most had not sought these experiences deliberately; they occurred spontaneously. But some people made a practice of going to special places where they felt inspired, like hilltops, meadows, lakes, woods, or beaches.

As a child, I spent a lot of time outdoors and felt a strong connection with the natural world, and a sense of belonging. This made me want to study science, especially biology. I was good at science at school, but what I learned was no longer based on direct experience of an organism's life. Virtually all the plants and animals we studied at school, and later at university, were dead. We killed what we studied, except for the animals used in vivi-section experiments, which were killed afterwards. We dissected earthworms, frogs, dogfish and rabbits. We took flowers apart and looked at their organs. We looked at tissues under the microscope. Killing animals for the sake of science was called sacrificing. The animals were being sacrificed on the altar of science.

This kind of science had very little relation to my own experi-ence. I tried to dismiss my own subjective feelings about the life of the natural world as unscientific, but they would not go away. I later came to realise that many people experience nature as truly alive when they are children, and they are encouraged to do so through children's stories and books about talking animals. As a child, I lived in an animistic world that was encouraged by adults. But as I grew up, it was made very clear to me that this childish way of thinking should be left behind. To believe that animals and plants are more than complex machines, and to think of nature as alive, was like believing in fairies.

Our entire culture is split between the experience of direct connections with the natural world, often established in childhood, and the mechanistic theory of nature that dominates the sciences and secular society. We are all inheritors of this split. In the official world, from 9 a.m. to 5 p.m. on Mondays to Fridays – the world of work, education, business and politics – nature is conceived of mechanistically, as an inanimate source of raw materials to be exploited for economic development. By contrast, in our unofficial private world, nature is identified with the countryside, as opposed to the city, and above all by unspoilt wilderness.[26]

Since the nineteenth century, and still today, many people want to get rich, if necessary by exploiting natural resources, so that they can afford to buy a place in the country to 'get away from it all'. On Friday evenings, the roads out of the cities of the Western world are clogged with traffic as millions of people try to get back to nature in a car. They are strongly motivated to do so. They are expressing a fundamental need.

How nature was split from God

How did this split come about?

One of its roots lies in the relationship between the Jewish people and the Holy Land in which they lived. The pre-Jewish religions of Palestine were polytheistic, with goddesses as well as gods, and they recognised many sacred places, including trees, groves, standing stones, mountains, springs and rivers. In the early stages of their life in the Holy Land, the Jewish people continued to worship at the ancient sacred places.[27] Things began to change with the building of King Solomon's Temple in Jerusalem, which was followed by attempts to suppress all other shrines, giving the temple a monopoly. The one God had one centre. Worship on hilltops, in sacred groves and at other ancient holy places was treated with suspicion, if not violent hostility.

The pre-Christian religions of Europe, like the pre-Jewish religions of Palestine, were polytheistic, and there were many holy places. But unlike the Jewish prophets and kings who tried to focus all ritual worship in one sacred place, Christians did not impose a monopoly. During the conversion of the Near East and Europe from the worship of the old gods and goddesses, many of the traditional sacred places and seasonal festivals continued in a Christianised form. In the Celtic Church in Ireland and Britain, local saints achieved a remarkable harmony between the druidic past and the new religion, like St Cuthbert (c.634–87), who was Abbot of the monastery of the holy island of Lindisfarne, but preferred to live as a hermit. According to the Venerable Bede (672–735), the author of *The Life and Miracles of St Cuthbert*, Cuthbert foretold many future events, and described 'what things were going on elsewhere.' He also spent his nights in the sea. According to a monk who crept out at night to observe him secretly:

When he left the monastery, he went down to the sea, which flows beneath, and going into it, until the water reached his neck and arms, spent the night in praising God. When the dawn of day approached, he came out of the water, and, falling on his knees, began to pray again. Whilst he was doing this, two otters came up from the sea, and, lying down before him on the sand, breathed upon his feet, and wiped them with their hair after which, having received his blessing, they returned to their native element. Cuthbert himself returned home in time to join in the accustomed hymns with the other brethren.

Sometimes practices of the old religions were assimilated as a matter of deliberate papal policy.[28] And to the old sacred places were added new ones connected with the saints: places where they had seen visions, where they had lived and died, and where their relics were enshrined.[29] I discuss this further in the context of pilgrimage in Chapter Seven. The incorporation of archaic religious elements into the Christian religion is still obvious in Roman Catholic and Orthodox countries. Think of the holy wells in Ireland, or the sacred mountain Croagh Patrick, or the many shrines of the Holy Mother of God.

Meanwhile, in early Christian theology, and in the orthodox teachings of the Middle Ages based on Aristotle and St Thomas Aquinas, nature was alive. The sun and the planets, the earth, plants and animals were all animated by souls. The living God was the source of this living world, and continually interacted with it. As the twelfth-century mystic, composer and Abbess St Hildegard of Bingen put it, 'The Word is living, being, spirit, all verdant greening, all creativity. This Word manifests itself in every creature.'[30] Medieval Christian theology was animistic and God's being underlay the being of nature. God was in nature and nature was in God.[31] Nature was alive, not inanimate and mechanical. The God of medieval Christianity, which gave us the great cathedrals of Europe, was the God of a living world.

In the sixteenth century, the Protestant Reformation led to a radical break in this Christian relationship to sacred times and places and to the natural world. The Reformers were trying to establish a purified form of Christianity, rejecting the corruptions and abuses of the Roman Church. Personal faith and repentance were what mattered; seasonal festivals, pilgrimages, devotion to the Holy Mother, and the cults of saints were denounced as pagan superstitions. As John Calvin put it, 'Nuns came in place of vestal virgins; the church of All Saints to succeed the Pantheon; against ceremonies were set ceremonies not much unlike.'[32]

The Reformers were trying to bring about an irreversible change in attitude, eradicating the traditional idea of spiritual power pervading the natural world, and especially present in sacred places and in spiritually charged material objects. They wanted to purify religion, and this purification involved the disenchantment of the world.[33] The spiritual realm was confined to human beings. By contrast, the natural world, governed by God's laws, was incapable of responding to human ceremonies, invocations, or rituals; it was spiritually neutral or indifferent, and could not transmit any spiritual power in or of itself.

The Protestant Reformation thus prepared the ground for the mechanistic revolution in science in the following century. Nature was already disenchanted and the material world was separated from the life of the spirit. The idea that the universe was a vast machine fitted this kind of Protestant theology, and so did the constriction of the realm of the soul to a small region of the human brain. The domains of science and religion could henceforth be separated. Science took the whole of nature for its province, including the human body; religion took the moral and spiritual aspects of the human soul.

With the seventeenth-century revolution in science, nature became machine-like, unconscious, inanimate, and lacking any purpose of its own. The world-machine was created and set in motion by God in the first place, but thereafter worked automatically. God's main role was in the supernatural realm, the realm of angels, spirits and human minds, but he still interacted with the realm of nature occasionally, by suspending the laws of nature and intervening through miracles. Isaac Newton thought that planetary orbits required occasional adjustments by supernatural intervention.

By the end of the eighteenth century, celestial mechanics had become more sophisticated. Theoreticians no longer needed miraculous adjustments to the machinery, and God became superfluous for understanding the workings of nature. His role was increasingly confined to the beginning and the end of time. At the beginning he was the creator; at the end he was the Judge at the Last Judgement. His religious role was primarily moral.

Thus God became increasingly remote. By the eighteenth century, many influential Enlightenment intellectuals such as Voltaire in France, and Thomas Jefferson and Benjamin Franklin in America, adopted the philosophy of Deism, in which God's remoteness was made explicit. God had created the machinery of nature in the first place according to rational laws and designs, but could not respond to worship and prayer. He provided no basis for the regular practices of the Christian religion, or any other religion. From here it was a short step to atheism. By assuming that the universe was eternal and needed no creator, the residual God of Deism became redundant.

The Romantic reaction

To start with, the mechanistic vision of nature was portrayed as a matter for celebration. For the rationalists of the eighteenth century, nature was a rational system of order, most clearly reflected in the Newtonian motions of the celestial bodies. Nature was uniform, symmetrical, and harmonious. She could be known through reason; she was indeed the very basis of reason and aesthetic judgement:

> First follow Nature, and your judgement frame
> By her just standard, which is still the same:
> Unerring Nature, still divinely bright,
> One clear, unchang'd and universal light.

Alexander Pope, 1711

But by the end of the eighteenth century, nature came to be understood in an almost opposite sense. Powered by sometimes dark and disconcerting unknown modes of being, she was irregular, asymmetric, and inexhaustibly diverse. In England this change in fashion was expressed through landscape gardening. Instead of clipped and manicured formal gardens, the landscaper sought to imitate an ideal of natural wildness. One model for the new style was found in paintings of pastoral scenes; another was in Chinese gardening.

Attitudes to wild places themselves changed radically. To most of our ancestors, forests, mountains, and wildernesses were dangerous. In the seventeenth century, travellers frequently referred to mountains as 'terrible', 'hideous', and 'rough'.[34] Even at the end of the eighteenth century, most Europeans found wild, uncultivated wilderness totally unpleasing: 'There are few who do not prefer the busy scenes of cultivation to the greatest of nature's rough productions,' wrote William Gilpin in 1791.[35] Dr Samuel Johnson shared the majority view, and said of the Scottish Highlands, 'An eye accustomed to flowery pastures and waving harvests is astonished and repelled by this wide extent of hopeless sterility.'[36]

The new taste for wild nature was sophisticated, inspired by literary and artistic models. Scenes were picturesque because they looked like pictures; they were romantic because they recalled the imaginary world of romances, far away and long ago.

By the beginning of the nineteenth century, the Romantic taste for wild nature led to dislike for human interference. The painter John Constable wrote in 1822: 'A gentleman's park is my aversion. It is not beauty because it is not nature.'[37] Romantic nature was best experienced in solitude, and part of the attraction of the wilderness was its remoteness from the bustle of cities and industrial activity. As travelling became easier, many well-off English people attached an unprecedented importance to visiting wild and romantic places. As Robert Southey wrote in 1807:

> Within the last thirty years, a taste for the picturesque has sprung up; and a course of summer travelling is now looked upon to be essential . . . While one of the flocks of fashion migrates to the sea-coast, another flies off to the mountains of Wales, to the lakes in the northern provinces, or to Scotland; . . . all to study the picturesque, a new science for which a new language has been formed, and for which the English have discovered a new sense in themselves, which assuredly was not possessed by their fathers.[38]

By the mid-nineteenth century, many people thought that solitude in natural surroundings was essential for the spiritual regeneration of city dwellers. Some wilderness should be preserved both for individuals and for the sanity of society as a whole.

Among English poets, William Wordsworth (1770–1850) was the most influential. Wordsworth repeatedly lamented the loss of connection with the divine, transcendent realm to which young children are open. His ode *Intimations of Immortality from Recollections of Early Childhood* begins with these words:

> *There was a time when meadow, grove, and stream,*
> *The earth, and every common sight,*
> *To me did seem*
> *Apparelled in celestial light . . .*
>
> *Heaven lies about us in our infancy!*
> *Shades of the prison-house begin to close*
> *Upon the growing Boy,*
> *But he beholds the light, and whence it flows,*
> *He sees it in his joy;*

The Youth, who daily farther from the East
Must travel, still is Nature's Priest,
And by the vision splendid
Is on his way attended;
At length the Man perceives it die away,
And fade into the light of common day.

In America, as in Europe, a Romantic sense of nature grew up under literary and artistic influences. In particular, Ralph Waldo Emerson's essay *Nature* in 1837 helped transmit a new vision of human relationship to the natural world. Instead of Americans trying to impose their own historically determined consciousness on the wilderness, they could recognise their true, living relation to the land. Emerson realised that this reverential attitude to nature was rare:

> To speak truly, few adult persons can see nature . . . The lover of nature is he whose inward and outward senses are truly adjusted to each other; who has retained the spirit of infancy even into the era of manhood . . . In the woods . . . a man casts off his years as the snake his slough, and at what period soever of life is always a child. In the woods is perpetual youth. Standing on the bare ground . . . the currents of the Universal Being circulate through me; I am part or parcel of God.[39]

By the 1850s, the opening of railways and the acceleration of economic development had made the uncolonised lands of America increasingly accessible. The wilderness could no longer be taken for granted. Henry David Thoreau, a disciple of Emerson's, was one of the first to sense the threat to virgin nature. He proposed, in vain, that each town in Massachusetts should save a five-hundred-acre piece of woodland that would remain forever wild. The greatest of the Emersonian lovers and defenders of wild nature was John Muir (1838–1914), founder of the Sierra Club and chief protector of Yosemite National Park (see Chapter Four).

A central feature of Romanticism was its rejection of mechanical metaphors. Nature was alive, and organic, rather than dead and mechanical. The poet Percy Shelley (1792–1822) was a romantic atheist, against religion rather than anti-spiritual; he had no doubt about a living power in nature, which he called the Soul of the Universe, or the All-sufficing Power, or the Spirit of Nature. He was also a pioneering campaigner for vegetarianism, because he valued animals as sentient beings.[40]

There were also Romantic Deists, who included the leading pioneers of evolutionary theory. Charles Darwin's grandfather, Erasmus Darwin, suggested that God endued life or nature with a creative capacity that was thereafter expressed without the need for divine guidance or intervention. In his book *Zoonomia*, in 1794, he asked rhetorically:

> Would it be too bold to imagine that all warm-blooded animals have arisen from one living filament, which the great First Cause endued with animality, with the power of acquiring new parts, attended with new propensities, directed by irritations, sensations, volitions and associations, and thus possessing the faculty of continuing to improve by its own inherent activity, and of delivering down these improvements by generation to its posterity, world without end![41]

For Erasmus Darwin, living beings were self-improving, and the results of the efforts of parents were inherited by their offspring. Likewise, Jean-Baptiste Lamarck in his *Zoological Philosophy* in 1809 suggested that animals developed new habits in response to their environment, and their adaptations were passed on to their descendants. A power inherent in life produced increasingly complex organisms, moving them up a ladder of progress. Lamarck attributed the origin of the power of life to 'the Supreme Author', who created 'an order of things which gave existence successively to all that we see.'[42] Like Erasmus Darwin, he was a Romantic Deist. So was Robert Chambers, who popularised the idea of progressive evolution in his best-selling *Vestiges of the Natural History of Creation*, published anonymously in 1844. He argued that everything in nature is progressing to a higher state as a result of a God-given 'law of creation'.[43] His work was controversial both from a religious and scientific point of view, but like Lamarck's theory it was attractive to atheists because it removed the need for a divine designer.

These different worldviews can be summarised as follows:

Worldview	God	Nature	Evolution
Medieval Christian	Interactive	Living organism	No
Early mechanistic	Interactive	Machine	No
Enlightenment Deism	Creator only	Machine	No
Romantic Deism	Creator only	Living organism	Yes
Romantic atheism	No God	Living organism	Yes
Materialism	No God	Machine	Yes
Panentheism	Interactive	Living organism	Yes

Some people identify their experience of more-than-human nature with God; others identify it with nature, as opposed to God, while others (myself included) see God in nature and nature in God, a worldview called panentheism.

The hidden goddesses of materialism

Nature is feminine. In Latin *natura* is a feminine noun meaning 'birth'. When personified, nature is Mother Nature. Many people who have a negative image of a Father God transfer their allegiance to Mother Nature. Instead of feeling connected to God the Father, they feel connected to the Great Mother. Others, including myself, see no need to choose between the Father and the Mother. The very use of these gendered metaphors implies that both are essential. Father and Mother are correlative terms: they both need the other.

If nature was the sole source of all life, and if life had evolved, then Mother Nature had to be credited with more and more freedom and creativity. Erasmus's grandson Charles Darwin succeeded in turning this Romantic vision of the creative power of nature from poetry into a scientific theory. Like the Romantics, Charles Darwin saw Mother Nature as the source of all forms of life. Through her prodigious fertility, her powers of spontaneous variation, and her powers of selection, she could create life without the need for the intelligent design of a machine-making God. With his customary honesty, he remarked: 'For brevity's sake I sometimes speak of natural selection as an intelligent power . . . I have, also, often personified the word Nature; for I have found it difficult to avoid this ambiguity.'[44] He advised his readers to forget the implications of these turns of phrase.

Instead, if we remember what the personification of Nature implies, we see her as the Mother from whose womb all life comes forth and to whom all life returns. She is prodigiously fertile, but she is also cruel and terrible, the devourer of her own offspring. Her fertility greatly impressed Darwin, but he made her destructive aspect the primary creative power. Natural selection, working by killing, was 'a power incessantly ready for action'.[45] In India, the black goddess Kali personifies this destructive aspect of the Great Mother.

For modern materialists, nature, or matter, is the source of all things: all life emerges from her, and to her all life returns. Indeed, the very word matter, on which materialism is based, comes from the Latin *materia*, from the word for mother, *mater*. Nature gives birth to us, encloses and contains us; she provides our nourishment, warmth, and protection, but we are utterly at her mercy, for she is also terrifying, uncaring and merciless; she devours and destroys. Materialism is not solely a philosophical theory. Below the surface, it is an unconscious cult of the Great Mother.

The recent revival of animism

In a new turn of the spiral, some atheist philosophers themselves are challenging the materialist theory of nature. This theory assumes that matter is unconscious and that it is the only reality – or more generally, in a variant of materialism called physicalism, the physical world is unconscious and is the only reality.

Starting with these assumptions, the very existence of human consciousness is almost impossible to explain. How can unconscious matter inside brains generate consciousness? Modern philosophers of mind call the very existence of human consciousness 'the hard problem'. Some dismiss consciousness as an 'epiphenomenon' of brain activity, rather like a shadow that does nothing. Others go so far as to deny that consciousness exists, or dismiss it as an illusion.[46] On the other hand, a minority take a traditional dualist view, treating matter and consciousness as totally different, seeing consciousness as immaterial and outside space and time. But then they have the problem of explaining how they are related to each other, how they interact.

An increasing number of philosophers, including the British philosopher Galen Strawson[47] and the American philosopher Thomas Nagel,[48] have come to the conclusion that there is only one way out of the materialist-dualist dilemma, namely *panpsychism*, the idea that even atoms and molecules have a primitive kind of mentality or experience. (The Greek word *pan* means all, and *psyche* means soul or mind.) Panpsychism does not mean that atoms are conscious in the sense that we are, but only that they have some aspects of mentality or experience. More complex forms of mind or experience emerge in more complex systems.[49]

These philosophers are not claiming that all material objects, like tables and cars, have minds or experiences or purposes. Tables and cars do not form, organise and maintain themselves, and they do not have purposes of their own; they are made by people in factories to serve human purposes. Only self-organising systems, in other words systems that form, organise and maintain themselves, have mind-like properties or experiences, including atoms, molecules, crystals, cells, plants and animals. And their mental aspects are not necessarily conscious. After all, much of our own mental activity is unconscious, which is why we speak of our 'unconscious minds'.

According to the panpsychist philosophy, in self-organising systems, complex forms of experience emerge spontaneously. These systems are at the same time physical (non-experiential) and experiential, in other words they have experiences.

As Strawson put it, 'Once upon a time there was relatively unorganised matter with both experiential and non-experiential fundamental features. It organised into increasingly complex forms, both experiential and non-experiential, by many processes including evolution by natural selection.'[50]

Unlike the usual materialist attempt to explain consciousness by saying that it emerges as an epiphenomenon or an illusion from totally unconscious matter, Strawson's and Nagel's proposals are that more complex forms of experience emerge from less complex ones. There is a difference of degree, but not of kind.

Panpsychism is not a new idea. It is another word for animism. Most people used to believe in it, and many still do. In medieval Europe, philosophers and theologians assumed without question that the world was full of animate beings. Plants and animals had souls, and stars and planets were governed by intelligences. But confusingly, Strawson regards panpsychism as an updated version of materialism. He is still an atheist, as is Nagel, and still thinks matter is the only reality, but he has broadened the definition of matter to include experience or mind. But this broadened, animistic materialism soon takes us far beyond the realm of old-school materialism.

If nature is alive, if the universe is more like an organism than a machine, then there must be self-organising systems with minds at all levels, including the earth, the solar system and the galaxy – and ultimately the entire cosmos. The more-than-human world includes all these levels of consciousness.

In lifting our attention from the earth to the sky, the most important of all the celestial bodies is the sun. The sun sustains all life on earth. If we take panpsychism seriously, then new questions inevitably arise. Is the sun alive? Is it conscious?

The conscious sun

As soon as you ask if the sun is conscious, you realise that you are violating a scientific taboo, the purpose of which is to stop us taking seriously what our ancestors believed. Throughout most of human history, most people thought of the sun as conscious. For some, like the Indians and classical Greeks, the sun was a god; for others, like the Japanese, a goddess. In northern Europe, the sun was also a goddess; in Latvian and Lithuanian mythology, she was called *Saulė*.

This mythological background is reflected in the gender of the words for sun. In Germanic languages, she is feminine – in modern German, *die Sonne*. In southern European mythology and in Latin-based languages he is masculine – in modern French, *le soleil*. Children implicitly think the sun is conscious, and draw it with a smiley face.

From the modern materialist point of view, the fact that people all over the world thought of the sun as alive, divine and conscious disqualifies this idea from serious consideration. It is nothing but a childish superstition, an animistic projection of human emotions onto inanimate objects. The fact that children think this way merely proves the point.

Nevertheless, since the early twentieth century, there has been an extraordinary rise of an unconscious sun-worshipping cult, the basis of a multi-billion-dollar tourist industry. Sun-drenched beaches have become mass tourist resorts, and people who go there are often called 'sun worshippers'. This aspect of modern life happens on holidays and at weekends; it is part of the Romantic side of our cultural divide.

The doctrine that the sun is unconscious has been built into science since the seventeenth century. The philosopher René Descartes *defined* matter as unconscious. He separated off consciousness into the realm of spirit, which he defined as immaterial. The immaterial realm consisted of God, angels and human minds. Everything else in nature, including the sun, the stars, the planets, the earth, all animals and human bodies, was mechanical and unconscious. The sun and other stars were unconscious by definition, and from a scientific point of view they have remained so ever since.

But if the universe is more like an organism than a machine, then so is our galaxy, and so is our sun. The sun has highly complex patterns of electromagnetic activity within it and on its surface. Its patterns of activity are much vaster and more complex than the electromagnetic activity in our brains. Most scientists believe that the electromagnetic activity within our brains is the interface between body and mind. Likewise, the complex electromagnetic patterns of activity in and around the sun could be the interface between its body and its mind.

Maybe the sun is conscious, and physical aspects of its mental activity are measurable, just as electrical patterns of activity in brains are measurable.

I cannot prove that the sun is conscious; but a sceptic cannot prove that it is unconscious. From a non-dogmatic point of view, the consciousness of the sun is an open question.

This question leads to more questions. If the sun is conscious, why not all stars? And if stars are conscious, then what about entire galaxies? Galaxies are complex electromagnetic systems, with vast electrical currents flowing through the plasma of the galactic arms, linked to magnetic lines of force millions of light-years long. The galactic centre may be like the brain of the galaxy, and the stars like cells in the body of the galaxy. There may be a vast galactic mind, far exceeding the scope of our sun's more limited mind, with vast electromagnetic extensions of its activity passing though the spiral galactic arms.

In 1997, I helped to organise a symposium called 'Is the sun conscious?' at Hazelwood House, in Devon, over the summer solstice.[51] We assembled a small group of people, including a cosmologist, a physicist, an expert on mythology, an Indian philosopher, and some psychologists. On the summer solstice itself, 21 June, we got up early and went to watch the sun rise over Dartmoor. It was cloudy and raining until the moment of sunrise, when the sun broke through the clouds, and a perfect rainbow appeared behind us.

If the sun is conscious, then what does it think about? What kinds of decisions can it make? We thought that one group of decisions might concern its wider body, the solar system. The sun's light permeates the solar system, and so does the solar wind – energetic streams of particles moving outwards from the sun. Fluctuations in solar activity change the intensity of the solar wind; they influence the Northern and Southern Lights, they affect the ionosphere and radio transmissions, and they modulate the frequency of lightning. When intense bursts of solar activity are pointed towards the earth, the huge outpourings of charged particles can cause power outages and catastrophic breakdowns of electromagnetic technologies. The American space agency NASA issues regular space weather forecasts so that we have some warning of solar events that affect our life on earth.[52]

If the sun is conscious and has control over its body, it could modulate the entire solar system, including life on earth, by choosing when and where to fire off solar flares and coronal mass ejections. The sun could, if it wanted to, close down our technologies by an ejection directed towards the earth, causing enormous power outages. We have set up long-distance power transmission systems, like the British National Grid, which can act like aerials for these solar pulses. A major outburst of solar activity could melt down transformers and entire power grids would fail; they might take months to repair. The sun also modulates life on earth more subtly, including influences from its eleven-year cycles in which sunspot activity rises and falls and the magnetic poles of the sun reverse.

The sun might also be concerned about its own peer group, the other stars within our galaxy, the Milky Way. We know almost nothing about interstellar communication, or in what way the entire galaxy modulates the stars. But this is another realm in which the sun's consciousness is likely to come into play.

We do not know what level of consciousness the sun has. Is it only about its own bodily functioning? Or is it more like a mind that knows what is going on within the solar system, including what we ourselves are doing right now? Perhaps the sun can sense directly what is happening on earth through the electromagnetic field. The electrical changes in our radio and TV transmissions, in our mobile phones and computers, in our brains and throughout our bodies are all within the electromagnetic field of the earth, which is embedded within the electromagnetic field of the sun, the ambient field in which everything on earth happens. If the sun's mind can sense what is happening through this field, then it could know what is happening on earth and everywhere else in the solar system.

Can we communicate with the sun? Certainly many people worship the sun, and make offerings and prayers to the sun, either in its own right or as a channel of God's light, or as both.

Among the practices of yoga is *surya namaskar*, salutation to the sun. I have done this practice almost every morning for more than forty years, which is one of the reasons I am so interested in the sun. Another Indian solar practice is the *Gayatri* mantra, which I have often chanted, a prayer for the divine light of the sun to illuminate our meditation.

From a spiritual point of view, the light of the sun is the light of the Spirit, which shines through the sun and all other stars. Except at sunrise or sunset, or seen through a light-absorbing barrier, this light dazzles and overpowers us. According to many spiritual writers, including St Anselm (c.1033–1109), the sun is like God and God is like the sun:

> Truly, O Lord, this is the unapproachable light in which thou dwellest; for truly there is nothing else which can penetrate this light, that it may see thee there . . . My understanding cannot reach that light, for it shines too bright. It does not comprehend it, nor does the eye of my soul endure to gaze upon it long. It is dazzled by the brightness, it is overcome by the greatness, it is overwhelmed by the infinity, it is dazzled by the largeness, of the light.[53]

Our new situation

We are in an unprecedented situation. The withdrawal of God, consciousness and purpose from the universe of mechanistic science has been accompanied by a vast expansion of our vision of nature in space and time, from the fleeting appearance of evanescent subatomic particles in the Large Hadron Collider to the discovery of trillions of galaxies beyond our own in a universe that has been evolving for more than 13 billion years. Now, with the re-emergence of panpsychism, this vastly expanded universe can take on a new life and meaning. Our direct experience of non-human nature can again lead us beyond our limited selves to a direct connection with the more-than-human world and the more-than-human consciousness that underlies it.

But before reaching for distant galaxies or the ultra-microscopic, it is best to start nearer home.

Two practices for reconnecting with more-than-human nature

A SIT SPOT

Find somewhere outdoors that you can sit quietly and safely, and where you can be alone. If you live near a wood, or meadow, or riverside, then find a sit spot in one of these places. Or find a sit spot very near where you live, even if it is in your garden, or on a roof. Unless you find a place nearby, you will find it too hard to visit and spend time there regularly.

The practice is simple. Be there. Get to know that place at different times of day and night, in different weather conditions, in different seasons. Be aware of the four directions, and the course the sun takes through the sky. Get to know the plants that grow there and the animals that live there and those that pass by. Listen to the wind. Listen to the birds and learn to identify them from their songs.

The tracker Jon Young points out that if you sit quietly for about twenty minutes, the animals around you will get used to you, and cease to experience you as a source of alarm. You will then be able to notice their alarm calls, especially those of birds. If you are in a garden their sounds may alert you to a cat walking nearby. In the woods, the alarms will tell you when a person or another animal is moving through the woods, and where they are.[54]

Through experiencing the life in your place, you will link your own life to the more-than-human world, and you will soon feel a greater sense of connection and belonging.

THE SUN

Greet the sun in the morning. Or if the weather is cloudy, turn towards the hidden sun. If your house or garden does not have an eastwards view, then acknowledge the daylight as it comes in through your window, flowing from the sun.

During the daytime, when there is an opportunity, turn towards the sun. Do not look at it directly. But at sunrise or sunset, when it is not too bright, look at it directly and give thanks for its light, and thank the source of all light, whose light shines through it. Ask for the divine splendour of the sun to illuminate your meditation.

4
Relating to Plants

My father, a herbalist and pharmacist, introduced me to many kinds of plants. As a child, my first job was weighing out herbs and putting them into packets. Thanks to my father, as a boy I could name most tree species and identify common wild flowers.

In addition, I was fortunate to have not just one, but two secret gardens. The first was from my earliest years. Near the centre of my hometown Newark-on-Trent, in Nottinghamshire, there was a large walled enclosure that contained about six gardens. We rented one of them, bounded on one side by a high stone wall between the garden and the street, and on the other sides by tall, thick hedges. Inside the garden was an orchard, giving us apples, pears and damsons. We also had raspberries and gooseberries, a vegetable garden and flowerbeds. A revolving wooden summer-house with a veranda rotated, when pushed, on circular rails, so that it could be turned to face the sun. The garden was full of butterflies and birdsong in the spring and summer. And unlike the garden around our house, this secret garden was remote from home concerns; it was another world, although only a few minutes' walk away.

I spent many hours there with my father and brother, and later on my own. I could work in the garden, play, look at plants, watch birds, read, or daydream. I often felt happy there. But then the land was needed for a playing field for a nearby school. The gardens were replaced by mown grass, bounded by wire-mesh fences.

Soon afterwards, a great aunt died and left us her house in Newark, which had a separate garden about a hundred yards away, again behind a high stone wall, and enclosed on the other three sides by fences and hedges. It was bigger than our previous garden, about half an acre, and had its own powerful calm. There was an orchard, fruit bushes, a vegetable garden, flowerbeds, and a large lawn that we used for tennis and croquet. I loved being in these gardens. I also spent many hours in woods and by streams and ponds in the countryside around Newark, roaming freely on my bicycle. My friends and I had a freedom that few children now enjoy.

As I say, I have always felt a close connection with plants. I studied science at school, and botany and biochemistry at Cambridge University, where I was awarded the university Botany Prize.[1] I spent ten years at Cambridge doing research on plant development. I worked on rainforest plants in Malaysia, where I was based in the Botany Department of the University of Malaya. From 1974 to 1985 I worked in an international agricultural institute in India, the International Crops Research Institute for the Semi-Arid Tropics (ICRISAT), near Hyderabad, where I was Principal Plant Physiologist. I am also a gardener. I have written dozens of scientific papers on plants.[2] In this chapter I restrict my discussion to flowers and trees.

Relating to plants seems to be a powerful human need, grounded in millions of years of evolutionary and cultural history. Most of the time, foraging or looking after plants is practical, mundane rather than transcendental. But plants can open a window to another way of being. Their beauty connects us with the richness and diversity of the natural world, and reminds us of the creativity of life. We can, if we like, try to persuade ourselves that this is all a matter of unconscious evolutionary mechanisms. But the direct experience of the forms of plants can take us beyond the realm of thought to a direct connection with the more-than-human world.

Flowers evolved about 100 million years before humans emerged.

Insects were the first to appreciate their beauty. But did this sense of beauty arise in simple nervous systems as an unconscious, mechanistic adaptation to the forces of natural selection? Or are animal minds and flowers tapping into a source of beauty that both pervades and transcends the natural world?

Flowers

Most people delight in flowers, and for millennia many cultures have assumed that gods, goddesses and God delight in them, too.

In Egypt, the pillars of many temples in Luxor are surmounted by carvings of lotus flowers. Despite the prohibition against the making of graven images in the Ten Commandments, the Temple of Solomon was decorated with carvings that included plants and flowers: 'And he carved all the walls of the house roundabout with carved figures of cherubims and palm trees and open flowers,' (1 Kings 6:29). Likewise, the prohibition of images in Islamic art does not extend to flowers, and many mosques, tombs and other sacred buildings are decorated with floral forms, including the Taj Mahal.

Many Buddhist temples contain images of the lotus flower, and the Buddha is often portrayed sitting on one. Lotuses and other flowers are taken to Buddhist temples as offerings. Hindus offer flowers to gods and goddesses as part of their worship in temples. Flowers are regularly used to decorate Christian churches. In the porches of many English village churches, one of the most prominent notices is the Flower Rota, showing who is due to provide and arrange the flowers for Sundays.

In his Sermon on the Mount, Jesus said, 'Consider the lilies of the field, how they grow: They toil not, neither do they spin; yet I say unto you that even Solomon in all his glory was not arrayed like one of these' (Matthew 6:28–29). These words are embedded in a series of sayings about trusting in God so that you can live in the present, rather than worrying about what is going to happen tomorrow.

Most scholars agree that Jesus is referring to wild flowers in general, not only to lilies. The point about wild flowers is that they grow by themselves, spontaneously, as they have been doing for millions of years. By contrast, most garden flowers are the result of plant breeding. They reveal hidden potentials in their wild ancestors. Although they are generally more spectacular than their progenitors, they are the product of selection by humans, and their cultivation demands human activity. Even in Jesus's time, field crops, fig trees and grapevines were already the result of thousands of years of human cultivation and selection.

One advantage of considering wild flowers is that they immediately take us outside the human world of work – of toiling and spinning – whereas cultivated flowers do not. As a gardener myself, I much appreciate the beauty of garden flowers, and spend hours cultivating them. But I find it easier to become absorbed in the contemplation of wild flowers, because in my garden I soon notice things I ought to be doing, like pulling up weeds. Weeds are plants in the wrong place, but wild flowers are in the right place; they do not make me think that I should be doing something.

To consider flowers is not purely to look at them passively. It helps to know something about them. As you begin to consider them, you realise that each species has its own kind of flower. When the Swedish naturalist Carl Linnaeus laid the foundations of the modern classification of plants in the eighteenth century, he realised that flowers provided a way of grouping plants into families. He classified plants according to what is called the sexual system, because flowers contain the plants' sex organs – the male anthers that produce pollen, and the female carpels that contain eggs. Around these sexual organs are the petals and the sepals.

This system is at first sight surprising, because it brings together plants that look very different. For instance, the pea and bean family, the Leguminosae or Fabaceae, contains plants with many different shapes of leaves, and a wide range of forms and sizes. Some are herbaceous annuals, like chickpeas; some are climbers, like runner beans; some are shrubs, like pigeon peas; and some are trees, like laburnums and acacias. But all have similar flowers, and all produce seeds in pods.

The structures of flowers fall into a small number of basic categories.

One large group of plants called the monocotyledons, or monocots for short, includes grasses, bamboos, palm trees, orchids, lilies, hyacinths, bluebells and irises. Despite their very different forms, their flowers have a basically three-fold pattern, with petals in threes, or multiples of three. For example, lilies have six petals, three plus three.

The other major group of flowering plants are called dicotyledons, or dicots. In some families, the flowers have a basic four-fold pattern, with petals in fours or multiples of four – such as the cabbage family, the Cruciferae or Brassicaceae, including wallflowers, mustards and cabbages. Other dicot families have a basic five-fold structure, with five petals, or multiples of five, such as the rose family, Rosaceae, which includes apples, strawberries and blackberries. And in some families the flowers are composites of many small flowers, which together make up a meta-flower, as in the family Compositae or Asteraceae, which includes daisies and sunflowers.

Fortunately, botanic gardens make it easy for anyone to recognise the variety of forms within a family, and also to see their family resemblances. Most of these gardens have what are called systematic beds, where plants of different species within the same family are planted together – in one bed members of the Fabaceae, in another, members of the peony family, the Peoniaceae, in another, members of the saxifrage family, the Saxifragaceae, and so on. Although these beds are cultivated, most of the plants within them were collected from the wild. I have spent many hours looking at these collections in my favourite botanic gardens: the Royal Botanic Garden at Kew, in West London, and the University Botanic Gardens at Cambridge and Oxford.

As Charles Darwin pointed out in *The Origin of Species*:

> If beautiful objects had been created solely for man's gratification, it ought to be shown that before man appeared there was less beauty on the face of the earth than since he came on the stage . . . Flowers rank amongst the most beautiful productions of nature; but they have been rendered conspicuous in contrast with the green leaves, and in consequence at the same time beautiful, so that they may be easily observed by insects. I have come to this conclusion from finding it an invariable rule that when a flower is fertilized by the wind it never has a gaily coloured corolla . . . Hence we may conclude that if insects had not been developed on the face of the earth, our plants would not have been decked with beautiful flowers, but would have produced only such poor flowers as we see on our fir, oak, nut and ash trees, on grasses, spinach, docks and nettles, which are all fertilized through the agency of the wind.[3]

Darwin was surely right. For the vast majority of their history, flowers had nothing to do with human beings. Flowers first evolved more than 100 million years ago, in the age of the dinosaurs. They must have evolved because insects and other animals enjoyed looking at them. The beauty of the flowers is dependent on animals' eyes, which means that animals must have an ability to appreciate colours and shapes. They must have a sense of beauty. How otherwise can the evolution of flowers be explained?

The evolution of the sense of beauty

Animals' sense of beauty may well have evolved first of all in relation to other animals of their own species. Most species show little interest in flowers, and members of the opposite sex are the primary focus of their aesthetic interest. Think of a peacock. The peacock's tail evolved long before humans arose on the earth. It exists because female peacocks find it beautiful, and because male peacocks compete for mates. It is a big disadvantage in other ways. When I was living in India, I saw a dog chase a peacock repeatedly; the peacock ran away and lumberingly took off, and for several days just managed to escape. But one day the dog managed to get a mouthful of its tail feathers. Peacock tails are a disadvantage when fleeing from dogs and other potential predators, but they work well in attracting mates.

Darwin called this phenomenon sexual selection, which he explained in utilitarian terms:

> I willingly admit that a great number of male animals, as all our most gorgeous birds, some fish, reptiles and mammals, and a host of magnificently coloured butterflies have been rendered beautiful for beauty's sake; but this has been effected through sexual selection, that is, by the more beautiful males having been continually preferred by the females, and not for the delight of man. So it is with the music of birds.[4]

But while Darwinian evolutionary theory may explain the survival value, it does not explain the sense of beauty itself.

What is this sense of beauty that humans share with many other animal species? Where does it come from? The answer that satisfies you depends on your starting point.

If you are a materialist, then you assume that the universe is unconscious. There is no purpose in it, no mind, and no appreciator of beauty except the neural mechanisms inside the brains of animals, which in the case of bees and butterflies are very small.

This then raises the question of the evolution of qualities. Are animals' responses to colours, shapes and smells no more than a result of random genetic mutations and natural selection? In the end, is the attraction to, or repulsion from, sensory stimuli nothing but a result of genetically programmed nerve mechanisms? Materialists answer 'yes' to these questions.

By contrast, if you believe that consciousness is inherent in nature you have a different starting point. From a panpsychist point of view, insects and other animals are aware, and have minds that are capable of appreciating beauty. Our own minds share in a sense of beauty widespread in the animal kingdom, and many of the forms and colours that appeal to other animals appeal to us, too. Flowers and many animals are beautiful for beauty's sake, as Darwin put it. *But what is beauty's sake?*

Is the source of beauty in nature alone, or does it transcend nature? Is there is a transcendent mind beyond time and space? In the tradition inspired by the Greek philosopher Plato, this ultimate reality contained the archetypal source of truth, goodness and beauty. All beauty in nature was derived from this transcendent mind, the ultimate source of all forms. In Christian theology, this transcendent mind is the mind of God. According to the Roman Catholic catechism, 'the manifold perfections of creatures – their truth, their beauty, their goodness – all reflect the infinite perfection of God.'[5]

Not just in the Roman Catholic but in most religious traditions, perhaps all, animal and human minds are ultimately derived from the consciousness that underlies the universe, and is present in all nature. An insect's mind is ultimately derived from the source of all consciousness, and so are our own minds in their very different way. The whole of nature is a reflection of the creative mind that underlies all things. This is a traditional Christian view, a traditional Muslim view, and a traditional Hindu view.[6] All the qualities we experience – colours, like green, and smells, like lavender – are present in the mind of God. We participate in the divine experience of qualities.

St Anselm, a mystic, theologian, and Archbishop of Canterbury from 1093 to 1109, thought of God as 'that than which nothing greater can be imagined'. God contains all conceivable possibilities. A God who could not experience the fragrance of a rose would be less than a God who could experience the fragrance of a rose. God includes the fragrance of roses, and all colours and all forms, and all other qualities experienced by humans and other animals. All qualities are in the divine mind.[7]

In this way of thinking, animal minds, including insect minds, are the underlying basis for the beauty of flowers, and their minds participate in the divine being. That is why they help bring forth the beauty of flowers. Their minds and their sense of beauty participate in God's nature, as the ultimate source of Truth, Beauty and Goodness. God is not like an engineer who creates a separate, mechanical world, as some mechanistic theologians suggested. God is within the natural world, in every part of it; and the natural world is within God, and participates in God's being and consciousness.

When Jesus suggested that we 'consider the lilies of the field', he invited us to live in the divine presence through wild flowers. Through wild flowers we can directly experience God in nature and nature in God.

Sacred groves and national parks

Long before people cultivated gardens, sacred groves retained something of the primal qualities of paradise. They still survive in many parts of the world. When land was being cleared for human settlement or agriculture, some areas were left undisturbed, and protected as sanctuaries for wildlife, and also for spirits, gods and goddesses. In India, there are still thousands of sacred groves, often surrounding temples or shrines. Some of them seem to be as old as the Indus Valley civilisation (3300–1700 BC). Others, in tribal areas, date from the time of the first tribal settlements in the area.[8] Many of these groves are rich in flora and fauna, and in some cases among the last refuges of endangered species. There are many such groves in Europe, too. In some cases, churches were built within them.

In the Holy Land, sacred groves, often on hilltops, provided sanctuaries 'shaded by the thick foliage of venerable trees', where, for many generations after the Israelites had settled in Palestine, the people 'resorted to offer sacrifice, and there, under the shade of ancient oaks or terebinths [a form of pistachio tree], their devotions were led by pious prophets and kings, not only without offence, but with an inward persuasion of the divine approbation and blessing,' as James Frazer put it in his fascinating book *Folk-Lore in the Old Testament*.[9] This worship still continued after the construction of the temple in Jerusalem in the tenth century BC. But by the seventh century BC, Hebrew prophets were denouncing worship in these groves, and trying to centralise the Jewish religion in the city temple.

My favourite sacred grove in Europe is the sanctuary of the Sainte-Baume, in the South of France. On the side of a large hill is a deep cave, where St Mary Magdalene is believed to have spent the last thirty years of her life. Beside the cave is a small monastery, and within the cave is a spring, the shrine of the saint, and an altar. Surrounding this grotto is an ancient deciduous forest, with beech, oak and other species creating a moist, cool, mossy microclimate, quite unlike the dry and scrubby vegetation of Provence that surrounds it. This sacred grove must long pre-date its role as a Christian sanctuary, but the fact that it was Christianised ensured its preservation, and it is still a major place of pilgrimage.

The minimal form of a sacred grove is a sacred tree. In India, many villages and temples contain sacred trees, often a pipal (*Ficus religiosa*), also known as the *bodhi* tree, the tree under which the Buddha was enlightened, or a banyan (*Ficus benghalensis*), another species of fig, which is the national tree of India. In Japan, cherry trees are the focus of *hanami* or flower-viewing when they are in blossom in the spring.

In the Holy Land, sacred trees included oaks and terebinths. The first recorded appearance of God to Abraham took place at an oracular oak or terebinth at Shechem, where Abraham built an altar (Genesis 12:6–9). Later, Abraham dwelt beside the oaks or terebinths of Mamre, where he built another altar (Genesis 13:18), and it was under these trees that God appeared to him and promised that his aged wife Sarah would bear him a son (Genesis 18:1–10). The Abrahamic faiths are rooted in sacred groves.

In Britain, oaks were one of the sacred trees of the pre-Christian druidic religion, and many ancient oaks are still treated reverentially. The longest-lived trees in Britain are yews, also sacred in druidic times, and the most venerable of these trees are found in churchyards. In the village of Compton Dundon, in Somerset, the great yew is some 1,700 years old. The church itself dates back some 750 years, so the yew would already have been almost a thousand years old when the church was built beside it.

The earliest Greek temples were made of wood, and the columns were trunks of trees. Later, when pillars were made of stone, in Corinthian columns the capitals at the top were carved with the representation of leaves. Roman, Byzantine and gothic columns were often topped by carved foliage, reminding us of the origins of columns as tree trunks. And in many gothic cathedrals, not only do the columns and vaulting above them recall sacred groves, but hidden among the carvings are mysterious Green Men, with faces made of foliage, or with leaves coming out of their mouths, like vegetation spirits.[10] The metaphor worked both ways: sacred groves as cathedrals, and cathedrals as sacred groves.

Although North America was desacralised by the Protestant settlers, by the nineteenth century an increasingly influential minority argued for the preservation of wild places as nature reserves. Sacred groves were reinvented. Henry David Thoreau (1817–1862), whom I have already mentioned, in his book *Walden* wrote about his intimate experiences of the woods and wildlife surrounding Walden Pond, in Massachusetts. He was convinced that 'we need the tonic of wildness'. As the nearby village of Concord expanded, and as the surrounding forests were cut down for farmland and firewood, he foresaw the need for conservation: 'Each town should have a park, or rather a primitive forest, of five hundred or a thousand acres, where a stick should never be cut for fuel, a common possession forever, for instruction and recreation.'

This idea was ignored in his lifetime, but he helped inspire the environmental movement in America, and gave a religious dimension to his outdoor experiences: 'My profession is always to be on the alert to find God in Nature, to know his lurking-places, to attend all the oratorios, the operas of nature.'[11]

John Muir (1838–1914) was a disciple of Thoreau, and he came to think of unspoiled nature as revealing the mind of God. He made a sharp distinction between nature and civilisation, and thought that 'wild is superior'.[12] Muir became a very effective activist for the preservation of wild places. He co-founded the Sierra Club and campaigned successfully for the establishment of the first national park, Yosemite, in 1890, and then for the Sequoia National Park and other wilderness areas. He was in effect the pioneer of the US National Parks system, and several parks are named after him, like Muir Woods and Muir Beach in California. The John Muir Trail, a 210-mile path in the Sierra Nevada mountain range of California, passes through several national parks and almost all of it lies within designated wilderness.

Both Thoreau and Muir explicitly compared the wild places they so admired to religious structures. In his *Journal*, Thoreau described 'the quiet and somewhat sombre aisles of a forest cathedral,' and Muir thought of Yosemite as 'a natural cathedral'. One of the mountains is actually named Cathedral Peak, and Muir described it as 'a majestic temple adorned with spires and pinnacles in regular cathedral style.'[13] For Muir, mountainous landscapes were God's best work: 'God himself seems to be always doing his best here, working like a man in a glow of enthusiasm.'[14]

The national parks were a new version of sacred groves on an unprecedented scale, and indeed Muir thought of them as reconsecrated sanctuaries. Yosemite was a paradise 'that makes even the loss of Eden seem insignificant.'[15] Now there are many national and local parks all over the US and in many other countries. Some of them are studied intensively by ecologists, and they are places where spiritual, aesthetic and scientific interests overlap. Some are very large, and some small. In Britain, some of the smaller conservation areas are officially designated as Sites of Special Scientific Interest (SSSI), of which there are more than 4,500.[16] Some are geological, intended to preserve specific geological features, but most are biological, some selected to preserve the habitats of particular groups of animals, such as newts, dragonflies or birds, others to conserve plant ecosystems that might otherwise be destroyed by development or farming, such as bogs, chalk grassland and woodland streams.

Thus sacred groves persist even in the modern secular world as national and local parks, wildlife sanctuaries, nature reserves, and sites of scientific interest. Many of us resort to them for inspiration and spiritual refreshment.

But we do not need to go to the countryside to find trees. There are many trees in urban gardens and parks, and many people, including me, like to spend time with them. Trees are often older than we are, and their presence puts our own lives into a perspective that we do not experience in our human-to-human interactions. Trees are literally greater than ourselves. They act as a bridge between the heavens and the earth, their roots in the ground, connected to the rich life of the soil, and interconnected with other plants through the living fungal network of mycorrhizae, with their branches reaching up into the sky and the sunlight, sensitive to every gust of wind, a home to birds and insects and many other living organisms. Trees connect us directly to the life of more-than-human nature.

Two practices with plants

FLOWERS

This practice requires a small and inexpensive piece of apparatus, a hand lens, or loupe. The best kind is also the cheapest, giving a magnification of times ten (x10).

To look through the lens, hold it up to one of your eyes so it is only about half an inch from the eye. The object you are looking at will be about an inch from the lens on the other side. You have to get up close. An easy way to practise is by looking at the skin on your arms, where you will see pores and hairs in much more detail than you have ever seen them before, unless you are an experienced hand-lens user.

This practice involves seeing flowers from a bee's-eye point of view. Bees can see flowers from a distance, and are attracted by them, some more than others. They land on them and crawl within them. As they do so, they enter an immersive colourscape. If you use your lens to move into the flower, you will enter new realms of experience from a bee's-eye point of view. This process works best with flowers that have tunnels through which bees pass, like nasturtiums and foxgloves.

A TREE

Get to know a particular tree, preferably one near where you live, so that you can visit it frequently and observe it at different times of day and night, and at different seasons of the year. If it is a flowering tree, make sure to spend time with it when it is in flower, at its moment of greatest glory. When you can, sit beneath it, and listen to the wind blowing through its leaves. Imagine its roots spreading out in the ground, at least as far out from the trunk as the branches spread, and probably much further. Its root system is linked to symbiotic fungal networks of mycorrhizae, which the plant feeds with sugars, and which in turn absorb minerals from the soil and pass them to the plant. The roots pull water out of the soil, the trunk reaches up into the sky, its leaves open to the sunlight. It is truly a connector of the realms of heaven and earth. If you hold its trunk, putting your arms around it, you can be aware of the sap rising in the wood and the sugars flowing downwards in the bark to feed the roots – a two-way flow. This is a flow reflecting the fundamental polarities of the tree, roots and shoots, and dark and light. Sap flows upwards from the darkness of the roots into the leaves and flowers, into the light. Sugars flows down from the leaves to feed the roots.

If you ask the tree a question, you may receive an answer, not in the form of a voice, but in what you see or feel or hear. If you are angry or upset, you can ask the tree to transform your emotions, absorbing your anger, or worry, or sorrow. Above all, if you form a relationship with a life form that is bigger and older than yourself, and which may go on living long after you die, it will help to put your life and your problems in a much broader perspective.

Appendix to Chapter Four: Family Orchards

Paradise, a place of timeless harmony, was a garden. In the Abrahamic faiths, it was the Garden of Eden. Humans lived in harmony with God and with the plants and animals around them until their fall from grace.

Gardens are still images or reflections of paradise, attempts to recreate something of this lost world of goodness and beauty, not least within cities. Beautiful gardens, both public and private, give joy to many millions of people.

At a more modest level, gardening is one of the most frequent ways of relating to the non-human world, and increasingly so. In the United States, in 2014, thirty-five per cent of all households were growing food at home or in a community garden.[17] In the UK, gardening is the most popular outdoor activity, involving about fifty per cent of the population.[18] In addition, many people keep houseplants, even if they do not have a garden. And many buy cut flowers for their homes.

In Britain, gardens of various sizes are usually attached to houses. In addition, many people have vegetable gardens at some distance from their house, in the form of allotments. But allotments are usually small, normally 16 or 32 to the acre, or 2,700 or 1,350 square feet. They are also very functional, and not at all secret.

I was exceptionally fortunate to have secret gardens with fruit trees when I was a child, and for years I thought of this as impossible in the modern world. Urban land capable of being developed is too expensive, and farmland usually changes hands in units of hundreds of acres. But I now think that family orchards are not only possible, but could become widely available.

Look at this situation from the point of view of a landowner. Imagine you own agricultural land near a town or city, and that this land is zoned for agriculture and cannot be developed for building. Imagine you take just one acre and divide it into about five gardens, each about one fifth of an acre, about 8,000 square feet. These would be large enough to contain an orchard, a vegetable garden, and grassy areas as well as flowerbeds. Such a garden could be a rough rectangular shape of 80 x 100 feet. The gardens could be surrounded by hedges, and laid out with access paths, a parking area for cars and bicycles, and maybe even a communal picnic area with a firepit for barbeques.

How much would these family orchards cost? In 2016, agricultural land in England cost up to £9,000 per acre. To make the maths simpler, assume the value of this acre to be £10,000. When divided into five, the land cost of each orchard would be about £2,000. Add the cost of putting in the pathways and hedges, laying out communal areas and perhaps installing hand pumps or connections to a water supply – say about £3,000 per orchard. Thus, each of these five family orchards would have a cost price of around £5,000. How much could they be sold for? I would guess at least £15,000. Maybe much more, if there were a low supply and high demand. In other words, this project would probably be profitable financially, and would not require grants or subsidies.

Farmers or landowners who already own land near towns might be reluctant to sell, but might instead want to rent or lease their orchards. How much might they fetch in rent? I would guess at least £20 a week each, or roughly £1,000 a year. And, depending on supply and demand, rental or lease values might be much higher. For comparison, average rental levels of arable agricultural land are currently about £100 per acre per year. An acre rented out as five orchards would bring in at least £5,000 a year, fifty times more.

Of course, there would have to be legal provisions that enabled orchard-holders to restrain their neighbours from using the orchard as a residence, or as a scrapyard, or as a place to make loud noises. There could be an orchard-holders' association with a joint share of common parts, like paths, parking areas, water supply and collective barbeque garden. In other words, the orchard complex could be run along the same lines as many blocks of flats.

In Russia during the Soviet era, and still today, millions of people had *dachas* outside the city where they would stay at weekends and in the summer, and use their gardens to grow fruit and vegetables and keep chickens. The area of a typical dacha is about 600 square metres (about 6,500 square feet), slightly less than the orchard size proposed here.

The family orchard scheme would differ from dachas in that these gardens could not, and should not, be used as residences, otherwise they would soon turn into housing developments.

The orchard scheme should not raise major difficulties for planning permission, because it does not involve building houses or structures for living in. Agricultural land would be turned into horticultural land, in both cases dedicated to the growing of plants.

Imagine what would happen if a trial scheme proved a success. The demand for family orchards would increase rapidly. There would be a strong incentive to supply them. And many families would then have orchards where their children could play, where they could grow their fruit and vegetables and enjoy an oasis of peace. There would be far greater biodiversity, because orchards with hedges, fruit trees, flowerbeds, and vegetable gardens contain many more kinds of plants and animals than a monoculture of an arable crop. All this is possible, feasible and desirable.

5
Rituals and the Presence of the Past

All human societies have rituals, including religious rituals and sacraments, national rituals, seasonal festivals, and rites of passage and initiation, like the rituals surrounding birth, marriage, and death. These are generally community based and follow a traditional, formal pattern. Rituals imply a kind of continuity, a memory transmitted from past generations to the present generation through the practice of the ritual. Is this just a matter of cultural inheritance and the blind following of precedents, or does it go deeper?

In this chapter, I discuss a range of ritual practices. I then show how they are illuminated by the hypothesis of morphic resonance, the idea that memory is inherent in nature. According to this hypothesis, all organisms, including humans, draw upon a collective memory and in turn contribute to it.

Origins, myths and rituals

In many societies, a kind of memory is presupposed in the myths on which the society is based. Myths are stories of origins. They concern the doings of gods, heroes, and superhuman beings. They propose that the reason why things are as they are, is because they were as they were. The present repeats the past. This repetition inevitably goes back to the first time something happened.

In our modern technological age, we are used to rapid changes, and people in all cultures are now aware of these changes, if only through the advent of smartphones. Everyone knows that they are part of a new world, unknown to their ancestors. Practically all governments in the world are committed to economic development through science and technology. The ideology of progress is an all-pervasive, modern orthodoxy. But in traditional societies, there was no such ideology. The present repeated the past. Even in modern societies, rituals are conservative and follow established forms.

An Australian anthropologist, Ted Strehlow (1908–78), who spent many years among the Northern Aranda Aborigines of Australia, summarised the basic principles as follows:

> The *gurra* ancestor hunts, kills, and eats bandicoots; and his sons are always engaged upon the same quest. The witchetty grub men of Lukara spend every day of their lives in digging up grubs from the roots of acacia trees . . . The *ragia* (wild plum tree) ancestor lives on the *ragia* berries which he is continually collecting into a large wooden vessel. The crayfish ancestor is always building fresh weirs across the course of the moving flood of water which he is pursuing; and he is forever engaged in spearing fish. If the myths gathered in the Northern Aranda area are treated collec- tively, a full and very detailed account will be found of all the occupations which are still practised in Central Australia.[1]

This idea of the past as a timeless model is alien to modern thinking, but in traditional societies all over the world, the mythic attitude predominated. The myths told stories of origins that took place in another time, the 'dream time', but which are still enacted in the present. Every technique, rule, and custom was followed because 'the ancestors taught it to us.'[2]

The purpose of many rituals is to connect participants with the original event that the ritual commemorates, and also to link them with all those who have participated in the custom in the past. Rituals cross time, bringing the past into the present.

In all cultures, the effectiveness of rituals is believed to depend on conforming to the patterns handed down by the ancestors. Rituals are traditional by their very nature. Gestures and actions should be done in the correct way; and ritual forms of language are conserved even when the language is no longer in everyday use. For example, the liturgy of the Coptic Church in Egypt is in the otherwise extinct language of ancient Egypt, the rituals of the Russian Orthodox Church are in old Slavonic, and the Brahminic rituals of India are in Sanskrit.

Rituals of remembrance

In rituals of remembrance, present participants are linked to the primal creative moment that the observance commemorates, and – again – to all those who have taken part in this ritual before them. The Jewish feast of Passover recalls the original Passover dinner on the night before the Jews began their journey out of bondage in Egypt through the wilderness and to the Promised Land. On that night, after a series of nine fearful curses, in the tenth and final curse on Egypt, the first-born sons of the Egyptians and their cattle were destroyed, while the Jewish people were passed over, because in each household they had sacrificed a male sheep or goat, and sprinkled its blood on the doorposts of their houses. They cooked and ate the sacrificial animal with bitter herbs and in haste, preparing for their departure the next morning. Through participating in this ritual, and by hearing the story that goes with it, present-day participants affirm their identity as Jews, and their connection with all the Jewish people who have participated in this tradition before, right back to the first Passover. They are also connected with all those who will come after them.

Likewise, the Christian Holy Communion connects participants with the original Last Supper of Jesus with his disciples, itself a Passover dinner, and with all those who have participated since.

This is the basis of the doctrine of the Communion of Saints. The sacred time of the Mass is connected to the many Masses that have preceded it, and it will in turn be connected to the Masses that follow it. In the words of Mircea Eliade, a historian of religion,

> '[the Mass] can also be looked upon as a continuation of all the Masses which have taken place from the moment when [the Mass] was first established until the present moment . . . What is true of time in Christian worship is equally true of time in all religions, in magic, in myth and in legend. A ritual does not merely repeat the one that came before it (itself a repetition of an archetype), but is linked to it and continues it.'[3]

Remembrance is at the heart of all Jewish liturgy and ritual. When Jesus at the Last Supper said, 'Do this in memory of me,' he was uttering a thoroughly Jewish statement. As the theologian Matthew Fox has pointed out, 'The essence of religion and the essence of ritual is healthy remembering. But it's not just about remembering human events like the Passover and the Exodus and human liberation. It is also about remembering the creation events – the new moon, the equinox, the solstice, the seasons.'[4] These religious rituals contain an element of remembering and re-enacting the Creation.

The principles of remembrance and hope for the future apply to many secular and national rituals as well. For example, the American Thanksgiving Day recalls the thanksgiving festival of the Pilgrim settlers of New England after their first harvest in 1621, itself in the tradition of thanksgiving days established in England by the Protestant Reformers, to replace a far larger number of Roman Catholic religious festivals.

Many other countries commemorate days when seminal historical events took place, or celebrate the birthday of the state. In France, Bastille Day on 14 July commemorates the storming of the Bastille, a fortress and prison in Paris, on that day in 1789, a turning point in the French Revolution. In India, Independence Day on 15 August commemorates the independence of India from the British Empire in 1947. Independence Day in Mexico, on 16 September, is celebrated with fireworks, parties and music to recall the 'cry of independence' on that day by Miguel Hidalgo, which helped trigger a revolt against the Spanish. In the Soviet Union, 9 November commemorated the 1917 Revolution that established the first communist government in Russia. In post-Soviet Russia, Victory Day, 9 May, recalls the victory over Nazi Germany in 1945. These secular, national rituals, like religious and tribal rituals, define the identity of those who participate in them, and connect participants with those who have gone before and with those who will come after.

Initiations and rites of passage

Initiation rituals are concerned with the crossing of boundaries such as those between boyhood and manhood, or between the unmarried and the married state. They are rites of passage. So are the rituals associated with the crossing of boundaries in space and time, from one country to another, or from one year to another. And so are the rituals of birth and death.

Through a study of a wide range of rites of passage in different cultures, the anthropologist Arnold van Gennep showed that typically they have three phases.[5] In the first, the initial state is removed. In rites of maturity, the state of childhood is stripped away; in many funeral customs, the person who has died is freed from the responsibilities of life: he or she no longer has the duties of a living person, and no longer needs to play the normal social roles. The individual is separated from his or her initial state and left in transition.

This threshold state is dangerous and ambiguous. In maturity rites, this state may be symbolised by going into the bush or forest far away from normal life, or by undergoing dangerous trials and ordeals. Finally, a ritual of integration ends this phase, and emphasises the individual's integration into his or her new state. Such rituals have many similarities across cultures. Washing, headshaving, circumcision and other bodily mutilations indicate separation, as do the crossing of streams and other obstacles, or spending time alone in the wilderness. Anointing, eating, and dressing in new clothes indicate integration.[6]

Initiation rituals carry individuals across social or religious boundaries, and at the same time they define these boundaries and make them manifest. The Gisu of Uganda say that they initiate boys to make them men so that they do not remain uninitiated boys. These rituals are not simply a way of marking biological maturity, because they are carried out with boys at different stages of maturity; they are concerned with the crossing of cultural boundaries. The initiations define the categories that they presuppose.

In traditional rites of passage for young men in some Native American groups, the initiates spent time alone in the wilderness, without food, water or shelter, and often in danger or pain. In this state of separation, they sought signs, dreams or visions that would help them and their community when they returned as men. These rites of passage are often called 'vision quests', and several organisations now guide people from the modern Western world through vision quests modelled on these traditional practices.[7]

Even many secular practices reflect features of initiation rites, such as the passing of tests and the awarding of certificates in schools; passing examinations and gaining university degrees in graduation ceremonies; inductions into professional bodies after passing professional tests; the commissioning of army officers after their military training, and so on.

I was in the first age group that was not conscripted for National Service in the UK armed forces following the Second World War.

For many of the young men who were older than me, being in the armed forces was a rite of passage. When I went to Cambridge, about half the undergraduates in my year had just finished their National Service, and many had been involved in active combat or served in places of mortal danger – in Malaya, or Kenya, or Cyprus. They had been at risk of losing their lives, not symbolically, but actually. In most modern societies, there are no longer rites of passage in which young people confront death. But they are continually reinvented. Gangs often have dangerous rites of passage for new gang members, involving trials by ordeal.

One reason why many young people in modern societies take psychedelic drugs is because they serve as a rite of passage. Bad trips can be terrifying, and some drugs induce a near-death experience (NDE), especially dimethyltryptamine (DMT),[8] one of the most intense of all psychedelics. But these initiations are often unguided, unlike traditional rites of passage, and can be dangerous and disorienting without a ritual of reintegration. In traditional societies, adolescent boys who have been through a rite of passage from boyhood to manhood are typically welcomed into the circle of initiated men. Likewise, girls who have undergone a rite of passage connected with their entering sexual maturity are welcomed into the circle of women. The same still happens in religious rites of passage such as the Christian confirmation ceremony and the Jewish bar mitzvah, or bat mitzvah for young women.

But people who take psychedelic drugs and undergo a transformative experience cannot be welcomed or reintegrated into the wider society because the drugs are illegal in most countries, and many people disapprove of them. However, there are now several religious groups, such as the Santo Daime Church in Brazil, a legal psychedelic Christian church, where the taking of a psychedelic brew – in this case *ayahuasca* – happens within a ritual, and participants are initiated, guided and helped by experienced elders.

(I plan to discuss the spiritual role of psychedelics in a sequel to this book.)

Near-death experiences and ritual drowning

Many people have had a near-death experience spontaneously; in fact, more do so than ever before, thanks to coronary resuscitation and modern medicine. Many who would have died in the past, now survive. There has been much research on this subject, and there are many books about NDEs, including the bestselling *Proof of Heaven: A Neurosurgeon's Journey into the Afterlife* by Eben Alexander, describing his own NDE when he was in a coma while suffering from meningitis.

Not everyone who nearly dies has an NDE; only a minority do so, about twelve to forty per cent; but that still means that many millions of people have had these experiences. And although most NDEs are highly pleasurable, a minority are not. Some people have the same kinds of experiences as positive NDEs, but they resist them and feel powerless, angry, or afraid. Other people feel they were completely alone in a void; and some find themselves in scenes of torment along with other human spirits in extreme distress.[9]

Many positive NDEs have essential features in common. They often begin with the experiencers floating out of their body, looking at their physical body from above, seeing themselves lying there with nurses and doctors attending them. Then they often go through a tunnel into the light, and feel that they are in a loving presence. They may meet deceased members of their family, or beings of light. Some experience a life review, when the events of their lives flash before them. Many describe this experience as blissful. But then, because it is a *near*-death experience and not a death experience, they are drawn back in their physical body. Some say that they have died and been born again.[10]

NDEs often bring about positive changes, including less fear of death, and more spiritual and loving attitudes. People who have experienced these changes often say that their NDE was the most profound and helpful experience of their life.[11]

Although everyone agrees that NDEs happen, their interpretation is hotly contested within the academic world. For materialists, it is inconceivable that consciousness can separate from the brain, and conscious life after bodily death is impossible, nothing but a superstition. Hence NDEs must be hallucinations, produced by the desperate activities of dying brains suffering from a lack of oxygen. However, some people have had NDEs while their brains were being monitored in operating theatres and showed no apparent electrical activity; they were 'flatlining'.[12] Nevertheless, materialists argue that precisely because they had these NDEs, there must have been brain activity to give rise to them, even if this activity was undetectable.

This is not really a scientific dispute about empirical facts, but a question of belief systems. For a diehard materialist, no amount of evidence will ever show that conscious experience is separable from brains, because it would contradict the materialist philosophy. By contrast, religious believers usually welcome this evidence.

Some traditional rites of passage may well involve NDEs, and current scientific research on NDEs can shed much light on these rituals. For instance, the NDE phenomenon enables us to reinterpret a key practice of initiation described in the New Testament, and practised in the early Church: baptism by total immersion. The prototypical baptiser was John the Baptist.

What if John the Baptist was a drowner? He baptised people by immersing them in the River Jordan, including Jesus himself. What if John held the initiates under for just long enough for them to experience an NDE by drowning? When they recovered from this near-drowning experience, many of them would have said that they had died and been born again; that they had seen the light; and they had lost the fear of death. This would have been a cheap, simple, rapid and effective way to induce a life-transforming experience of death and rebirth. I imagine that people who had been suitably prepared would have lined up on the banks of the Jordan, and that John would have baptised one after another, with helpers to aid their recovery. They may have lost a few. But that was before the days of Health and Safety laws and liability litigation.

Everything that the New Testament says about the experience of baptism makes sense if the people being baptised were having NDEs. The alternative is to argue that this experience of death and rebirth was symbolic. But to be symbolic, it would have to be symbolic of a near-death experience by drowning. Why do something symbolic when people could experience the real thing?

Early Christians practised adult baptism in the tradition of John the Baptist, but infant baptism was already widespread by the second century AD, and in the third century was standard practice, although adults could still be baptised by having water poured over their heads three times, as they can today.

One surprising aspect of the Protestant Reformation was a revival of baptism by total immersion. In the religious ferment in sixteenth-century England and Germany, groups of radical reformers reinstated the baptism of adults by total immersion. They were called Anabaptists: the Greek prefix *ana* means 'again' or 'back to'. The early Anabaptism movement gave rise to a range of churches and religious communities including the Mennonites and the modern Baptist Churches, which still practise the baptism of teenagers and adults by total immersion.

I suspect that in the sixteenth and seventeenth centuries, the Anabaptists rediscovered this experience of an NDE through drowning. Even today, of all Christian denominations, those who talk most about the experience of death and rebirth – being born again – are Baptists. Perhaps few modern Baptists have NDEs during their baptism because of the modern concern with Health and Safety, but in previous centuries, the importance of a direct experience of death and rebirth may have outweighed the fear of going too far.

A grotesque perversion of this ritual of death and rebirth is the use of waterboarding by the US intelligence services. This is a form of water torture in which the victim is tied down on a sloping board at an angle of about ten to twenty degrees, face upwards, with the feet higher than the head. The victim's face is covered with a cloth, and water is poured onto the face, causing a gag reflex and sensation of drowning.

Ironically, this form of torture was invented by the Spanish Inquisition in the sixteenth century to deal with Anabaptists, who were persecuted both by Roman Catholics and by mainstream Protestants as heretics. The Anabaptists believed in an adult baptism, since they rejected the value of infant baptism. In 1527, King Ferdinand of Spain declared that death by drowning, which he called a 'third baptism' was the proper response to Anabaptism.[13] As a recent article in a theological journal by William Schweiker explained:

In the Inquisition, the practice was not drowning as such, but the threat of drowning, and, symbolically we can say, the threat of baptism. The *tortura del agua* or *toca* entailed, like waterboarding, forcing the victim to ingest water poured into a cloth stuffed into the mouth in order to give the sense of drowning . . . It was, we must surely say, a horrific inversion of the best spirit of Christian faith and symbolism. This poses questions . . . Is waterboarding a kind of forced conversion hidden within a political action and thereby all the more powerful as a tool in the hands of the state to demonize its enemy? Does this signal a breakthrough of the demonic within political and military action since a religious rite is being subverted for immoral ends? These questions are so buried in public discourse that their full import is hardly recognized, even by devout Christians.[14]

Sigmund Freud might have called this 'the return of the repressed'. In his book *Moses and Monotheism*, he wrote, 'What is forgotten is not extinguished but only "repressed".' What has been repressed does not 'enter consciousness smoothly and unaltered; it must always put up with distortions.'[15]

These distortions should not blind us to the fact that near-death experiences are transformative for most people who experience them. They are of great positive value.

According to the biblical accounts, an initiation through ritual drowning was at the root of Jesus's own experience of God as a loving father. His baptism by John the Baptist awakened him to his direct relationship to God. According to St Mark's gospel, 'And just as he was coming out of the water, he saw the heavens torn apart and the Spirit descending like a dove on him. And a voice came from heaven, "You are my Son, the Beloved, with you I am well pleased".' (Mark 1:10–11.)

Ritual sacrifice

Until recently, some traditional societies practised human sacrifice, which is now banned by law everywhere. But it still happens, and is now called ritual murder. In 2006, in the Khurja region of Uttar Pradesh, India, about eighty-five kilometres from Delhi, according to the local police there were dozens of child sacrifices to the goddess Kali.[16] In 2008, a rebel commander in Liberia's civil war admitted to taking part in human sacrifices as part of traditional ceremonies intended to ensure victory in battle. He said the sacrifices 'included the killing of an innocent child and plucking out the heart, which was divided in pieces for us to eat.'[17]

In many cases, animal sacrifice is explicitly recognised as a substitute for human sacrifice, as in the Old Testament story of Abraham and his son Isaac. Abraham, believing that God required him to sacrifice Isaac, was about to do so when an angel of God stopped him, and he sacrificed a ram instead (Genesis 22:2–8).

In the Jewish Passover story, when God was about to unleash the final and most dreadful of his ten curses on the Egyptians, killing their first-born sons and the first-born of their cattle (Exodus 11:4–6), the Jewish people were passed over because they did as Moses had told them: each household slaughtered a male lamb, and sprinkled or smeared its blood on the doorposts and above the door. The slaughter of the lamb acted as a substitute for death of Jewish young men and cattle. The Jewish people also had a ceremony in which all the sins of the community were laid upon a goat, which was then driven to its death in the wilderness taking their sins away with it. This was the original scapegoat (Leviticus 16:8).

From a modern secular perspective, the idea that a person or an animal should be sacrificed to save others makes no sense. But in evolutionary terms, it is a deep-seated pattern. When predators such as lions attack a herd of animals, they identify one member of the group who seems especially vulnerable, because it is young, or old, or lame, and kill it. When they have done so and their appetite is satisfied, the other members of the herd relax; they are safe for a while. The death of one member of the group has saved the others.[18] The same theme underlies stories of dragons that threaten whole communities, and who can be appeased by being offered a child, often a virgin girl, as a victim. She dies for the sake of the others; she saves them by her death.

In her book *Blood Rites*, Barbara Ehrenreich argues persuasively that for most of human history, humans were more like scavengers than hunters, and lived in continual fear of predation:

> Humans, and before them, hominids, could not always have been the self-confident predators depicted in the standard museum diorama. The savannah that our hominid ancestors strode (or,

more likely, crept warily) into was populated not only by edible ungulates, but by a host of deadly predators, including a variety of big saber-tooth cats as well as the ancestors of lions, leopards and cheetahs. Before, and well into, the age of man-the-hunter, there would have been man-the-hunted.[19]

Walter Burkert, a historian of religion, imagined the 'unritualised, real situation' from which sacrificial rituals arose in terms of:

> a group surrounded by predators: men chased by wolves, or apes in the presence of leopards . . . Usually there will be but one way of salvation: one member of the group must fall prey to the hungry carnivores, then the rest will be safe for the time being. An outsider, an invalid, or a young animal will be most liable to become the victim.[20]

To this day, primates like chimpanzees are often victims of predation. In a recent study of a forest chimpanzee population, predation by leopards was the principal cause of death, and lions were also significant killers. Troops of savannah-based baboons are often attacked and some lose a quarter of their members to predation every year.[21] When they move through the savannah, they fall into a defensive marching order with the young males on the periphery. A sick baboon that falls behind tries so hard to catch up that it exhausts itself, and is soon the victim of a predator. Sometimes the young males literally sacrifice themselves in the defence of the group; as a result, among wild primates, a high proportion of young males do not survive to maturity.[22]

These are not merely facts of life for wild animals, or archaic ways of thinking in primitive societies. They are alive and well today. Every soldier, sailor or member of an aircrew is a potential victim, prepared to die to save other members of their nation. In the twentieth century, at least twenty million young men did so in the First and Second World Wars. And today many members of armed forces, insurgent groups, freedom fighters, jihadists, and suicide bombers lay down their lives for others, in the ultimate self-sacrifice. They are heroes and martyrs for the people they are trying to save. The rhetoric of sacrifice helps provide a motivation for the risks they take, and also for the way their deaths are recognised and appreciated by the groups they are fighting for.

For many secular-minded people, the most baffling aspect of Christianity is its portrayal of Jesus saving others through his death on the cross. And indeed, it makes no sense without the whole historical tradition of sacrifice in general, and in Jewish history in particular. How can Jesus be like a sacrificial lamb and take away sins? He is like the sacrificial lamb at Passover, and also like the scapegoat. The annual Jewish ritual of offloading the sins of the community onto a literal scapegoat, and driving it out in the wilderness to its death was one ingredient in this Christian imagery. In the Mass, just after the consecration of the bread and wine as the body and blood of Jesus, the *Agnus Dei* is sung or said:

> O *Lamb of God, that takest away the sins of the world, have mercy upon us.*
> O *Lamb of God, that takest away the sins of the world, have mercy upon us.*
> O *Lamb of God, that takest away the sins of the world, grant us thy peace.*

Jesus's sacrificial death only makes sense in the context of animal sacrifice that is found in many religions, including the Jewish religion, which provided the historical context for the Christian interpretation of Jesus's death. A male sheep was a substitute for human sacrifice in the story of Abraham and Isaac; and the death of male lambs protected first-born Jewish sons in the Passover.

The killing of male sheep substituted for the sacrifice of first-born male humans. But Jesus reversed this process. The sacrifice of a first-born male human substituted for male sheep and goats, ending animal sacrifice.

Animal sacrifice continues to this day in Judaism and Islam. Jews still slaughter lambs at Passover, and Muslims sacrifice cows, sheep, goats or camels on Eid al-Adha, also known as Bakr-Eid, in commemoration of Abraham's sacrifice of a ram instead of his son Isaac. By contrast, for Christians, the crucifixion of Jesus reversed and terminated this process. Animal sacrifice was superseded by a full and final human sacrifice, the sacrifice of Jesus on the cross.

Sacrifice on the altar of science

Although the idea of substitutionary sacrifice seems nonsensical from a modern, secular point of view, it now happens on an unprecedented scale. This sacrifice does not take place in public, like traditional religious sacrifices, but behind closed doors in scientific laboratories. Within the US alone, about twenty-five million vertebrate animals are killed each year in biomedical research, mainly mice, rats, birds, zebra fish, rabbits, guinea pigs and frogs, with smaller numbers of dogs, cats, monkeys and chimpanzees.[23] They are sacrificed on the altar of science for the good of humanity. Indeed, the technical term for the killing of these animal victims is 'sacrifice'. A search on Google Scholar for scientific papers containing the phrase 'rats were sacrificed' brings up about 68,000 results, and 'mice were sacrificed' about 108,000.[24]

For biology students, sacrificing their first animals is a kind of rite of passage, just as dissecting a human corpse is a rite of passage for medical students. Sacrifice and dissection necessitate a dissociation from normal human sentiments and emotions; those who are initiated are supposed to adopt a persona of scientific detachment. A young scientist, Alison Christy, reflected on her experiences in an eloquent blog as follows:

> The first time I worked with rodents, I was a high school student doing a neuroscience research project at the University of South Alabama. In order to get clear brain histology, we had to perfuse the animals with saline. This means that the rat, a big, white animal, was injected with some kind of anesthetic, and we watched it run around a plastic tub until it became loopy and clumsy and finally lay still. We then placed it on a board and drove pins through its paws, crucifixion-style. We looped a string over his front teeth to hold his head back. We took shiny thin scissors and cut into the animal's skin and right through his ribcage. Inside the ribcage was the dark-red, still-beating heart. Blood starts to clot in the brain as soon as an animal dies. To get clean slices of brain, we had to push the blood out while the animal was still alive . . . To perfuse an animal with saline, you insert a needle into the left ventricle of the still-beating heart, and you cut the right atrium of the heart with scissors. Then you push your solution through . . . Quickly, the liver blanches and the paws, nose and tail become pale. The animal is entirely bloodless. The brain will be free of contaminating blood.[25]

As Christy pointed out, most researchers get used to these kinds of procedures. She noticed that the same thing happened with medical students as they got used to dissecting corpses. To start with, they were silent and serious, and some even fainted or vomited. 'But a week later you'll see them chatting and laughing with their lab partners as they wiggle their fingers around the vessels of the heart. A few months later they'll dissect the face without any hesitation . . . They act like completely different people than the ones who entered the lab only months before.'

Secular humanists reject the idea that humans can be saved by

God, and instead put their trust in human science and reason. Many see scientists themselves in the role of saviours, liberating humanity from ignorance and suffering. But the old archetype of sacrifice has not gone away; science itself relies on it. Nor have fears of destruction through cataclysmic events been banished by expelling gods and goddesses from the secular world. As well as the threats to human survival created by climate change and environmental destruction, we live under the shadow of vast arsenals of nuclear bombs that could yet unleash the ultimate holocaust.

In its original meaning, a holocaust was an animal sacrifice totally consumed by fire (Greek *holo* = whole; *kaustos* = burnt). But whereas in the ancient world, relatively small numbers of animals were burnt on altars as offerings, a modern scientific holocaust through weapons of mass destruction would consume millions of people, and countless animals as well. Sacrificed for what or to whom? Not to God or to Mother Earth, but as a display of human power. Collectively we are potentially more vengeful and terrifying than any of the vengeful gods or goddesses in myths and legends. Our potential for acts of destruction is vaster. And stories of divine destruction were located safely in the past. The scientific holocaust is located unsafely in the future, and we do not know whether it can be avoided or not.

Collective and individual rituals

All religions have collective acts of worship and thanksgiving, especially on holy days and festivals. All religions have ceremonies for marriage, deaths and the naming of babies. Secular humanists recognise the need for such ceremonies and have constructed rituals of their own.

Many people practise rituals within their families, such as saying grace before meals. Some practise rituals individually, in private prayer and meditation, and through yoga, chi gong, tai chi and other spiritual disciplines.

Everyday life contains many more or less unconscious ritual elements, such as shaking hands. Convention decrees that this should be done with right hands rather than left hands. Stone carvings from ancient Greece show that this custom goes back to at least the fifth century BC. Handshaking may have begun as a gesture of peace, demonstrating that the right hand held no weapons. In the modern world, it is part of a brief ritual of greeting, parting, making agreements or congratulating.

Many of us bless each other when parting, even if we are unconscious of what we are doing. The word *goodbye* is a mutated and shortened form of the blessing, 'God be with you'. *Farewell*, originally 'Fare thee well', is also a blessing. *Adieu*, a French form of goodbye, is literally *à dieu*, meaning 'to God', with the implied meaning 'I commend you to God'. In Spanish, *adiós* means the same. Other ritualised words of parting contain an implicit prayer for preservation until meeting again: 'see you' – in French, *au revoir*; in German, *auf Wiedersehen*; in Italian, *arrivederci*.

Rituals are part of all our lives. We cannot live without them. But we have a choice over the rituals we take part in, and the spirit in which we do so. They can be dull and habitual. Or they can be enlivening, inspiring and spiritually rewarding.

Morphic resonance

Why is the effectiveness of rituals so widely believed to depend on their similarity to the way they have been done before?

The way in which we understand rituals depends on our assumptions about the essence of nature. Ritual activities are related to deep-seated ideas of how minds and nature work. They make much more sense if nature, societies and minds contain a kind of memory, and less sense if they do not.

The usual assumption in science is that the basic ordering principles of nature, the so-called laws of nature, are fixed.[26] They were already present, fully formed, like a cosmic Napoleonic Code, at the moment of the Big Bang, when our universe came into being. Stars, atoms, molecules, crystals and living organisms behave as they do because they are governed by these eternal laws, which are the same at all times and in all places.

This assumption was grounded in sixteenth- and seventeenth-century theology, when the founders of modern science – Copernicus, Kepler, Galileo, Descartes, Boyle, Newton and others – assumed that nature was governed by the *logos*, the eternal mind of God. The eternal mathematical laws of nature were ideas in God's timeless mind. That is why they were invisible and immaterial, yet present everywhere. They shared in God's immutable, omnipresent and omnipotent nature.[27]

Eternal laws made sense in the context of a non-evolutionary worldview and a non-evolutionary theology. But our cosmology is now radically evolutionary, and many scientists reject the idea of an immaterial, all-pervasive mind that sustains the laws of nature. Nevertheless, eternal laws remain the default scientific assumption, because most scientists think there is no alternative. But since the beginning of the twentieth century, some philosophers and scientists have suggested that the laws of nature might evolve, just as human laws evolve. Or, to use a less anthropomorphic metaphor, the so-called laws of nature may be more like habits. Memory may be inherent in nature. Stars, atoms, molecules, crystals and living organisms may behave as they do because their predecessors behaved that way before. Each biological species may have a collective memory on which each individual draws and to which it contributes. Instincts may be like habits of the species. A young orb-web spider may know how to spin its web without being taught because it has inherited the memory of web-spinning from countless previous spiders.

My own hypothesis is that nature's habit-memory works through a process I call morphic resonance, which involves the influence of like upon like across space and time. Similar patterns of activity or vibration pick up what has happened in similar patterns before.[28] The more often a pattern of activity has occurred, the more likely it is to occur again, other things being equal. The more the repetition, the deeper the grooves of habit. When habits are very deep-seated, like the behaviour of hydrogen atoms or nitrogen molecules, they look as if they are changeless, as if they are governed by eternal laws. If we consider only long-established phenomena, it is impossible to tell the difference between eternal laws and long-established habits, because in both cases the same phenomena occur in much the same way over and over again. The difference between these two interpretations becomes experimentally observable when we consider new phenomena that have never happened before.

For example, when chemists make a new chemical compound and crystallise it, according to the eternal-law theory, it should crystallise the same way on the first, thousandth and billionth occasion, because the relevant laws of quantum theory, electromagnetism, thermodynamics, and so on, are always and everywhere the same. By contrast, if habits build up in nature, the substance may be very hard to crystallise for the first time, because there is not yet a habit for that kind of crystal to form. But the more often these crystals are made, the easier it should be for crystals to form all around the world as a new habit builds up.

By morphic resonance, the second time the crystals are made, they should form more readily because of an influence from the first crystals, other things being equal; the third time more readily still, because of an influence from the first and second crystals; the fourth time yet more readily, because of morphic resonance from the first, second and third crystals, and so on. Eventually this cumulative memory will lead to their crystallisation following a deep groove of habit, and the rate of crystallisation will reach an upper limit.

What actually happens? It is in fact well known that the more often crystals are made, the more readily they tend to form elsewhere. Turanose, a kind of sugar, was considered to be a liquid for decades before it first crystallised in the 1920s. Thereafter it formed crystals all over the world.[29] Reviewing cases such as this, the American chemist C.P. Saylor commented that it was 'as though the seeds of crystallisation, as dust, had been carried upon the winds from end to end of the earth.'[30]

There is no doubt that small fragments of previous crystals can act as 'seeds' or 'nuclei' that facilitate the process of crystallisation from a supersaturated solution. That is why chemists assume that the spread of new crystallisation processes depends on the transfer of seeds from laboratory to laboratory, like a kind of infection. Thus the formation of new kinds of crystals provides one way of testing the hypothesis of morphic resonance.[31] Increased rates of crystallisation should still be observable, even if visiting chemists are kept out of the laboratory and dust particles are filtered out of the air.

The hypothesis also applies to behaviour. If rats in London learn a new trick, rats all around the world should be able to learn it quicker, simply because the rats have learnt it there. The more that learn it, the easier it should get elsewhere. There is already evidence from experiments with laboratory rats that this remarkable effect occurs.[32] Likewise, it should be easier for people to learn what other people have already learnt, and there is scientific evidence that this is so.[33]

The key to morphic resonance is similarity. Its usual effect is to reinforce similarities, leading to the build-up of habits. By contrast, rituals involve the reverse of this process. In rituals, patterns of activity are deliberately and consciously performed the way they were done before. In habits, previous patterns are repeated unconsciously; in rituals, they are repeated consciously. In habits, the presence of the past is unconscious; in rituals, it is conscious.

Through morphic resonance, rituals bring the past into the present. The greater the similarity between the present ritual and the past, the stronger the resonant connection.[34] Thus morphic resonance provides a natural explanation for the repetitive quality of rituals found in traditions all over the world, and illuminates the way in which rituals connect present participants with all those who have done the ritual before, right back to the first time it was performed.

But rituals are not only about connecting across time; they are about opening to the spiritual realm in the present, just as people opened to this realm in the past. Repeating the same actions will help to bring about the same kind of spiritual connection. Americans taking part in the festival of Thanksgiving are giving thanks to God in the present, as well as linking up with previous generations of Americans giving thanks to God.

Two ways of participating in rituals

RITUALS OF GREETING AND PARTING

We can become more conscious of our greeting and parting rituals. When we shake hands, we can see this as a gesture of peace. When we kiss or hug, we can become aware that this physical connection has ancient biological and social roots. Apes such as bonobos frequently kiss each other, and dogs and cats lick and nuzzle one another as expressions of intimacy and trust. Some animals exchange food by passing it from mouth to mouth, as adult wolves do with their cubs, and in some cultures, human mothers pass chewed-up food to their babies directly from mouth to mouth. Of course, kisses can be erotic, but in many cultures, they have long played a much more widespread social role in greetings and part-ings, including among the ancient Persians, Egyptians, Jews, Greeks and Romans.[35] Early Christians exchanged the 'kiss of peace', and in contemporary Roman Catholic and Anglican church services, members of the congregation exchange a 'sign of peace', a kiss, hug or handshake, as part of the communion liturgy.

The expression of peaceful intentions is explicit in many forms of greeting, as in the Muslim greeting *assalaamu alaikum*, peace be with you, and the similar Jewish greeting *shalom aleichem*, peace be upon you. For Hindus, saying *namaste* or *namaskar*, meaning 'I bow to you,' together with the *anjali mudra*, the gesture of placing the hands together, can mean 'I bow to the divine in you.' In all cases, these rituals can be treated as mere conventions, but they take on a new power and significance when we become more conscious of their deeper meanings.

Likewise, rituals of parting can take on a greater meaning and power when we recognise the blessing that is implicit or explicit in them, as in 'goodbye', 'adieu', 'adios' and 'God bless you'.

CHORAL EVENSONG

Choral Evensong is an Anglican evening service in which choral music is sung, psalms chanted, and ancient poetry and prayers recited. On weekdays, it usually lasts around forty-five minutes, and on Sundays about an hour, because on Sundays it includes a sermon. Choirs in cathedrals, abbeys, churches and chapels have sung Evensong ever since the time of Queen Elizabeth I in the sixteenth century. This is one of the great cultural and religious treasures of the Anglican Church, with its beautiful sixteenth-century English, and wealth of musical settings. Great Elizabethan musicians such as Thomas Tallis and William Byrd created exquisite polyphonic music for this service, and new musical settings have been composed ever since.

Choral Evensong takes place in hundreds of churches and cathedrals every Sunday evening, not just in Britain, but in Ireland, the United States, Canada, Australia, New Zealand and other parts of the English-speaking world. In many cathedrals, abbeys and college chapels it happens on weekdays as well, when highly trained choirs sing extraordinarily beautiful music echoing around these great sacred spaces. This service is often candlelit and is followed by the playing of the organ. In Roman Catholic cathedrals and monasteries, there is a similar evening service called Choral Vespers.

These contemplative, restful, peace-inducing services are open free of charge to all. If you are a Christian or come from a Christian background, much of the language will resonate with your own experience and that of your ancestral tradition. If you are an atheist or agnostic, you will probably find Choral Evensong inspiring and uplifting. And if you are from a different religious tradition, this service will give you a taste of the Christian tradition and provide an easily accessible way of participating in it. People of all faiths and none are welcomed. For Britain and Ireland, a website, www.choralevensong.org, provides information on where to find Choral Evensong and Choral Vespers, when they will happen, and details of the choirs and the music they will sing.

Attending Choral Evensong or Choral Vespers provides a simple way of taking part in a long-established ritual that gives a strong sense of continuity over time, and which can confer blessings on all who share in it.

6

Singing, Chanting and the Power of Music

I have been immersed in music since I was born. But I claim no special distinction. Almost all humans throughout all human history have been immersed in music. In all traditional societies, singing and dancing are part of the group's collective life. Music plays a part in all religious traditions. Even in modern secular societies, music pervades most homes through radios, televisions and sound systems, and is present in many public spaces, even if only as background music in shopping centres and hotels.

My mother played the piano, my father the flute, and my paternal grandfather was a church organist and choirmaster, as was one of his sons, my father's younger brother. At the age of five, I began learning the piano, and the organ at age fifteen. I sang and chanted in the choir at my Anglican preparatory school, and later at my Anglican secondary school. As an undergraduate at Cambridge, I sang in a madrigal choir. By that time I was an atheist, and did not attend church services regularly. Nevertheless, I enjoyed going to Choral Evensong. I also played the college organ.

Meanwhile, when I was an undergraduate, I stayed during a vacation with a friend who lived near Liverpool, where we first encountered the Beatles in the Cavern, just before they rocketed to stardom. They opened a new dimension of musical experience for me. The Rolling Stones came soon after.

When I was working and living in India, in Hyderabad, I often heard groups of Hindus in villages and temples singing *bhajans*, devotional songs to goddesses and gods, and ecstatic music at Sufi shrines. I also lived in a Christian ashram in Tamil Nadu, where I sang and chanted five times a day.

I first met my wife, Jill Purce, in India in 1982. Jill was then leading, and still leads, chanting and sound healing workshops. She is a pioneer of the revival of group chanting, drawing on many different cultural traditions, including Mongolian and Tuvan overtone chanting. In her workshops, she offers a powerful and direct experience of the fundamental principles of chanting shared by traditions all around the world.[1]

We all have our own musical biographies, and they are all different. As the neurologist Oliver Sacks put it, 'for virtually all of us music has great power, whether or not we seek it out or think of ourselves as particularly "musical". This propensity to music shows itself in infancy, is manifest and central in every culture, and probably goes back to the very beginnings of our species. Such "musicophilia" is a given in human nature.'[2]

In this chapter, I discuss the evolutionary origins of singing, chanting, and dancing; then their effects on people's wellbeing, the physiology of participants and the cohesion of groups. Then I look at music in the context of physics and consciousness, and end by asking why most cultures assume that gods, goddesses, angels, spirits and God like music. Is this purely a human projection? Or is it an insight into the nature of ultimate reality?

The evolution of singing and music

Songs do not leave fossils, so we have no concrete evidence of the vocal activities of our remote ancestors. But we can learn a lot by looking at other animal species, at fossils and archaeological remains, and by comparing human musical traditions.

Charles Darwin led the way in thinking about the evolution of music. In his book *The Descent of Man, and Selection in Relation to Sex*, he discussed 'the capacity and love for singing or music' in a wide range of animals. He pointed out that some species of insects and spiders produce rhythmic sounds, usually by rubbing together special structures on their legs. In most species, only males make these sounds. He thought that their chief purpose was 'to call or charm the opposite sex.' In some species of fish, males make sounds in the breeding season. Air-breathing vertebrates have a pipe for inhaling and expelling air, and hence have the potential for making sounds by modifying the flow of air through a vibrating organ. In the amphibia, most notably in frogs and toads, males croak and sing during the breeding season, sometimes in chorus. Some reptiles make sounds, as do many species of bird.

As Darwin put it, 'The male alone of the tortoise utters a noise, and this only during the season of love. Male alligators roar or bellow during the same season. Everyone knows how much birds use their vocal organs as a means of courtship; and some species likewise perform what may be called instrumental music.'[3] One example is the drumming sound produced mechanically by snipe through the vibration of their outer tail feathers when they dive through the air as part of their courtship display. And woodpeckers drum rather than sing to attract mates, pecking rapidly on resonant objects to create a characteristic pattern of sounds. Darwin also drew attention to species of mammals that make musical sounds, including singing mice and gibbons.

Not only do many species make sounds themselves, but they also seem to be attracted to music. Why? Darwin had no answer:

But if it be further asked why musical tones in a certain order and rhythm give man and other animals pleasure, we can no more give the reason than for the pleasantness of certain tastes and smells. That they do give pleasure of some kind to animals, we may infer from their being produced during the season of court-ship by many insects, spiders, fishes, amphibians, and birds; for unless the females were able to appreciate such sounds and were excited or charmed by them, the persevering efforts of the males, and the complex structures often possessed by them alone, would be useless; and this is impossible to believe.[4]

In most animal species, only males sing. But in some monkey and ape species, most notably gibbons, both sexes sing. And so do male and female humans.

Darwin thought that music had a very ancient origin, which would help to explain its presence in all human cultures. He pointed out that flutes made from reindeer bones and horns had been found in caves along with flint tools and the remains of extinct animals, suggesting that they had been made and used a very long time ago. The recent radiocarbon dating of bone pipes and flutes from caves in France and Germany has shown that the oldest were made about 40,000 years ago, soon after our species, *Homo sapiens*, arrived in Europe.[5] Ian Cross, a modern researcher on the evolution of music, thinks that their sophisticated design suggests that 'music was likely to have been of considerable importance to a people who had just come to inhabit a new and potentially threatening environment.'[6] Music may have helped these new settlers in Europe adapt to this unfamiliar and uncertain world by promoting bonding and greater group cohesion. And the use of musical instruments probably came long after the development of singing and dancing.

Darwin also drew attention to the importance of music in inducing emotions. He pointed out that in oratory, musical elements are used to stir feelings in the audience. 'Even monkeys express strong feelings in different tones – anger and impatience by low, fear and pain by high notes.'[7] He also argued that the long evolutionary history of responses to music would help to account for its effects on the emotions:

We may assume that musical tones and rhythm were used by our half-human ancestors, during the season of courtship, when animals of all kinds are excited not only by love, but by the strong passions of jealousy, rivalry, and triumph. From the deeply laid principle of inherited associations, musical tones in this case would be likely to call up vaguely and indefinitely the strong emotions of a long-past age.

In this context, as Darwin observed, it is surely relevant that 'love is still the commonest theme of our songs.'[8]

Darwin also suggested that the evolution of singing and language were closely related. He thought that singing came first, and that speech evolved from music. In this regard, he anticipated much modern evolutionary thinking about the origin of language. However, rather than music preceding language, the musicologist Steven Brown has proposed that they both arose from a common communicative system, 'musilanguage'. When they diverged, language became more important for exact communication, and music played a predominantly social role, to do with the bonding and unity of the group.[9]

Social entrainment

The fossil evidence suggests that the capacity for making proto-musical sounds could have evolved as long as 1.8 million years ago in *Homo ergaster* and *Homo erectus*, who both walked upright and had brain sizes around 1,000 cubic centimeters (cc), not much less than the modern average of 1,400 cc. Their barrel-shaped chests and enhanced vocal capacities, together with ear canals similar to those in modern humans, suggest that the sounds of voices were already of great importance for their social lives. By 700,000 years ago, with the emergence of *Homo heidelbergensis*, a fully modern vocal tract appeared, together with ears that were maximally sensitive to sounds in the range of speech and song.[10]

No one knows when human societies first discovered the power of synchronised movement and sound-making. Non-human primates do not have the ability to sing together with a steady beat, although chimpanzees and bonobos sometimes make short bursts of synchronised calls.[11] As soon as proto-humans developed this ability, singing and dancing probably arose together. Through coordinating their sounds and movements, they discovered a power whereby the whole was more than the sum of the parts. This synchronised activity would have had huge effects on the members of the group itself, and also on other species. Predators would have been impressed by a display of united group power.[12]

Even today, hikers in Canada and the United States are taught to respond to bears, cougars and other threatening predators by trying to make themselves look bigger than they are, by raising their hands and making loud noises. If this works with one person, it must work better with ten people stamping, moving their arms, and chanting in synchrony. It would also have impressed other human groups. Many tribal societies used war chants, which still exist in a domesticated form in football chants, as in the Maori *haka* or war chant performed by the New Zealand All Blacks rugby team before they begin a match.

Humans have a very long history of mutual entrainment. Even in modern cities this tendency emerges spontaneously and unconsciously. When people are walking together and chatting, they often entrain each other without thinking about it, and come into step.[13] Our natural tendency to walk in step is formalised in military marching. When troops march, they move more cohesively and efficiently than if they just stroll along in random groups. This principle played an important part in military discipline two thousand years ago in the Roman Army, and modern armies still put on impressive displays of group power through marching columns of soldiers, accompanied by drumming and martial music.

The most widespread forms of mutual entrainment take place through chanting, singing and dancing. People breathe together, make sounds together, and move in synchrony. They come into a resonant, rhythmical relationship with the other members of the group. Even when people are not participating in making music or dancing to it, but are sitting down as part of an audience, they are still entrained in a suppressed way, and many move or beat time with the music.

Although Darwin was surely right about the competitive role of music in courtship, he neglected the cooperative role of music in human societies, which is now the dominant theme in discussions of musical evolution.[14] In traditional societies, music is primarily participatory. Everyone takes part through singing or dancing, or both. Through musical participation, people take on a group identity, and experience and express emotions together.[15] In most cultures music is an essential component of rituals, including rites of passage, weddings, funerary rites and seasonal festivals.[16] Music both helps to maintain group cohesion, and is an expression of it.

Thus, in an evolutionary perspective, music probably emerged both in the context of courtship and sexual competition, as Darwin suggested, and also as an expression of group solidarity, connectedness and unity. Through taking part in the same songs, dances and chants, people often feel themselves part of a greater whole. In traditional dances, they are linked to all those who have done the same dances before them, and sung the same chants.[17] According to the hypothesis of morphic resonance (discussed in Chapter Five), they resonate with ancestral dancers and singers, bringing the past into the present.

Chanting

All scholars of the evolution of music agree that vocal music preceded instrumental music, as it does in our own individual lives. Many mothers sing to their babies or talk to them in a cooing way, sometimes called 'motherese', and many young children learn to sing nursery rhymes. If they learn musical instruments, they usually do so after they have already learned to sing.

Chanting differs from singing primarily in that it is more repetitive. A short phrase can be sung to a simple tune over and over again, as in the chanting of Hindu and Buddhist mantras. Or else a simple melody can be sung repeatedly with different words, as in the chanting of prayers and psalms in Eastern Orthodox, Roman Catholic and Anglican liturgies. In chants, unlike songs, there is usually no fixed rhythmic beat; they often follow the rhythm of the words.

Much of what I know about chanting comes from my wife, Jill Purce,[18] who, as I say, has been giving voice workshops and teaching chanting for over forty years. She pioneered a way of teaching how to experience the power of the voice that is common to all spiritual traditions. In her workshops she shows how chanting, especially repetitive chanting, brings groups of people into a literal resonance with each other. Through chanting mantras, the whole group can enter into a kind of resonance with those who have chanted the same chants before (discussed below). Here are some of the basic principles about vowels and mantras that she teaches.

Vowels

Compared with speaking, chanting and singing both involve a lengthening of the vowel sounds. Vowels are produced with the mouth open, and a continuous flow of air from the lungs. Consonants interrupt the flow of air; they are articulated by blocking it (p, b, t, d, k, g), diverting it through the nose (n, m), or obstructing it (f, v, s, z).

Even when chanted on the same note, the different vowels sound different. This is because they have different patterns of harmonics or overtones, caused by different shapes within the throat, mouth and modulated by the position of the tongue. These vowel sounds set up specific patterns of vibration, not only in the vocal organs, but also within other parts of the body.

You can experience this for yourself. Block your ears by putting your fingertips in them. Then try chanting on a single note the vowels *ee* (as in *sheet*), *eh* (as in *bed*), *ah* (as in *car*), *oh* (as in *got*) and *oo* (as in *boot*). The more effectively you block your ears, the more you will experience the vibrations internally, located in different parts of your body. For example, when you do this, you find that the *ah* sound is located primarily in your chest, and the *ee* sound in your head, vibrating your skull – so this literally vibrates your brain within it.

The m and n consonants also have vibratory effects that you can experience by blocking your ears and humming *mm* and *nn*.

Mantras

Mantras are sacred sounds, often in ancient languages, like Sanskrit. There are certain sounds, or particular strings of sounds, for specific illnesses or other circumstances, some for entering a state of clarity and emptiness, and some for tuning into a lineage of teachers.[19]

Mantra-like chants are used in many traditions, including among the Sufi mystics of Islam, where forms of chanting are combined with rhythmic movement of the body and rhythmic breathing, which can help bring the chanters into a state of ecstasy or bliss.[20] Through chanting together, breathing and making the same sounds people come into synchrony with each other.[21]

Some mantras are exoteric, widely known and an essential part of regular prayer and ritual practice. Others are esoteric, transmitted from teacher to student in private.

The best-known and most fundamental mantra in the Indian tradition is ॐ, *Om*, or *Aum*. The sounds are spread out roughly like this: *aa-oo-mm*. You can explore its immediate physical effects by blocking up your ears, and chanting it on a single note. When I do this, I feel the *aa* sound mainly vibrating in my chest. If I then move towards the *oo* sound via a short *o*, (as in *or*), the vibrations move upwards towards the throat, and then as I move to the *oo*, they sound in the lower part of my head, and the *mm* sound sets up vibrations spreading out from my nose.

The most fundamental mantra in the Christian, Jewish and Islamic traditions is *Amen*, rather similar to *Om*. But although it is spelled Amen in the Latin transliteration, its original pronunciation in the Jewish tradition, in the Eastern Orthodox Churches and in Islam is *Ameen*. (In the Greek New Testament it is written αμην, where the second syllable, η, eeta, is a long e, as opposed to ε, epsilon, a short e. Latin has only one letter for e.)

The original form, αμην, is more powerful mantrically than the Latin transliteration. Try both versions for yourself. When I do it, in the Latin form, both the *aa* and the *eh* vowels are in my chest, and the vibration leaps to my nose region on the *m* and the *n*. Chanting *aa-mee-nn* has a very different effect. After the vibration of the nasal vibration of the *mm*, the *ee* sound resonates with the outer part of my skull before the resonant centre shifts back to the nasal region in the *nn*.

Exoteric mantras are widely known and used, but esoteric mantras are more specialised, and have been passed down over many generations from teacher to student, transmitted orally. Tibetans travel great distances for the transmission of a mantra from a teacher. When people use these mantras, in Hindu and Tibetan traditions they believe, as I say, that they are tuning in to the whole lineage of teachers. This connects them with those teachers' attainments, their states of spiritual connection with ultimate consciousness.

This is a point where the traditional understanding of mantras and my own ideas about morphic resonance converge. When people are chanting a mantra together in a group, they are simultaneously resonating in at least three ways: first, with physical resonances within their vocal tracts and bones, as discussed above; secondly, through the resonant entrainment of the members of the group with each other, chanting the same sounds in synchrony to a shared pulse; and thirdly, through morphic resonance between those chanting in the present and all the people who have chanted the same mantra in the past, tuning in across time.[22]

Effects of singing together

One advantage of repetitive chanting, or of singing simple songs in unison, is that everyone can join in, even if they think that they do not have a good voice or cannot sing in tune. No doubt this experience of connection and unity is a major reason for the use of chanting and singing in practically all traditional societies, communities and religions. And it is probably one of the main reasons why so many people join church choirs or community choirs in the modern world. These are voluntary activities, and people would not take part unless they derived some benefit. And indeed scientific surveys of people who sang in choirs found that most said that singing together made them feel better, and contributed to their mental and emotional wellbeing.[23]

These subjective impressions are also accompanied by measurable physiological changes. Saliva samples taken from participants before and after singing showed significant increases in immunoglobulin A (s-IgA), pointing to an enhanced activity of the immune system.[24] This form of immunoglobulin is secreted externally into body fluids, including mucus in the bronchial, genital and digestive tracts, and is a first line of defence against microbial infections. In one study with a classical choir, s-IgA levels increased on average by 150 per cent during rehearsals and 240 per cent during the performance.[25]

Studies with residents of nursing homes who sang together

showed significant reductions in standardised measures of stress and depression, compared with those who did not sing. In one year-long study, independent elderly people who sang in a community choir showed significant improvements in physical and mental health.[26] In patients with dementia, both singing and listening to music alleviated some of their troubling symptoms, including depression and agitated and aggressive behaviour.[27]

After an extensive review of research on the effects of choral singing, one group of researchers summarised their conclusions as follows:

- Choral singing engenders happiness and raised spirits, counteracting feelings of sadness and depression.
- Singing involves focused concentration, which reduces worrying.
- Singing involves deep controlled breathing, which counteracts anxiety.
- Choral singing offers a sense of social support and friendship, which ameliorates feelings of isolation and loneliness.
- Choral singing involves education and learning, which keeps the mind active and counteracts decline of cognitive functions.
- Choral singing involves a regular commitment to attend rehearsal, which motivates people to avoid being physically inactive.[28]

The beneficial effects of music are so well established that they form the basis of a therapeutic profession, music therapy, which can be used to help adults or children with behavioural or emotional disorders, for pain management, relaxation, and in many other contexts, including therapy with pregnant women and their unborn babies.[29]

From the age of about three months, the foetus can hear sounds and responds to music in the womb, as shown by its movements, revealed by scans. And babies and infants are often calmed by music, which is why mothers in many cultures have always sung lullabies to them.

Not surprisingly, stimulating music has stimulating physiological effects, and tends to increase heart rate, breathing rate and blood pressure, partly through activating the release of adrenaline; slow music is associated with decreases in these measures. These physiological changes are controlled by activity within the brain stem, the part of the brain that joins on to the spinal cord, through which motor and sensory nerves pass from the main part of the brain to the rest of the body. The musical tempo affects the firing of nerve cells in the brain stem, bringing them into synchrony with the music.[30] A similar synchronisation occurs in the cerebellum, which is concerned with the coordination of movements and balance.[31] Both the brain stem and the cerebellum are evolutionarily ancient parts of the brain, within the so-called reptilian brain.

The effects of different kinds of sounds may be related to ancient evolutionary instincts:

[M]usic commonly classified as 'stimulating' mimics sounds in nature, such as the alarm calls of many species, that signal potentially important events (e.g., loud sounds with sudden onset and a repeating short motif). Interestingly, positive affect and reward anticipation have also been associated with high frequency, short motif calls. This, in turn, heightens sympathetic arousal (heart rate, pulse, skin conductance, and breathing). By contrast, 'relaxing' music mimics soothing natural sounds such as maternal vocalizations, purring and cooing (soft, low-pitched sounds with a gradual amplitude envelope), which decrease sympathetic arousal.[32]

Whereas rhythms are primarily linked to the brain stem and cerebellum, melodies are primarily processed in the right hemisphere of the cerebral cortex, the opposite side of the brain from the primary language-processing areas.[33] And not surprisingly, pleasurable music also activates regions of the brain (in the mesolimbic system) that are involved in arousal and the experience of pleasure.[34]

Another effect of music rooted in ancient evolutionary history is its effect on the levels of the hormone oxytocin, the so-called love hormone, which is found in many invertebrates and in all vertebrates, where it is produced in the brain and secreted from the pituitary gland. (Chemically, oxytocin is a peptide, made up of a string of nine amino acids.) This hormone is involved in reproductive behaviour and egg laying even in earthworms, and in the courtship and sexual activity of frogs and toads, reptiles and birds, where it stimulates bonding behaviour and singing.[35] The same is true of mammals, including singing mice and hamsters.[36]

Likewise, in humans, oxytocin plays a role in social bonding, sexual activity, and during childbirth. In breast-feeding mothers, the release of oxytocin into the bloodstream is part of the milk let-down reflex, which usually occurs when mothers hear their baby cry. Oxytocin levels within the brain cannot be measured directly, but its concentration in the blood increased in babies who heard their mothers singing to them in motherese. In other studies, oxytocin levels increased when people sang,[37] and patients in hospital after an operation who heard soothing music were more relaxed and had higher oxytocin levels than those who did not hear music.[38] Oxytocin facilitates trusting behaviour and reduces fear and anxiety.

Musical deprivation

As we have just seen, music has many positive effects on health, wellbeing, social bonding and group cohesion. The corollary must be that lack of music has negative effects on health, wellbeing, social bonding and group cohesion.

In her book *Dancing in the Streets: A History of Collective Joy*, Barbara Ehrenreich argues that musical deprivation is linked to an increase in the incidence of depression in modern secular societies, where few people sing together.[39]

In many tribal societies and hunter-gatherer communities, practically everyone sings and dances together. But as agricultural societies developed, with the growth of cities and social hierarchies, there was a conflict between ecstatic dancing and the preservation of social order. People in states of ecstasy have a reduced awareness of the surroundings and of normal social constraints. They are more open to altered states of consciousness, which can include a sense of spiritual connectedness and great joy. In hierarchical societies, the preservation of the dignity and authority of the higher-ups conflicts with their participation in dances with their social inferiors. In some societies, festivals have relieved this tension through allowing a reversal of the social order, as in the festival of Saturnalia in ancient Rome, on 17 December, when servants became masters, and masters servants.

I experienced this reversal of roles myself when I lived in India. While working at the International Crops Research Institute for the Semi-Arid Tropics, near Hyderabad, I lived in the wing of a crumbling palace. The palace was owned by a young raja, whose family were part of the traditional nobility of the Hyderabad State. The raja and his wife the rani were devout Hindus, and normally led a sedate life. On the eve of the festival of *Holi*, soon before the vernal equinox, they invited me to join them for the bonfire in the palace courtyard.

During my first year in India, I had no idea what to expect, and was amazed by what happened. I found myself in a gathering that included all the servants and their families. The dancing around the bonfire was wild. The young *mali*, or gardener, a lively young man, was now the master, and hurled insults at the raja, speaking to him in the most disrespectful and familiar forms of address, amidst peals of laughter. The next morning, the rani gave me a glass of a 'special Holi drink', which turned out to be *bhang*, a powerful cannabis concoction. With everyone high and in a fully festive mood, we ran around squirting coloured water at each other. Again, there were no distinctions of class or caste. Everyone had fun. The next day normal life began again, but it felt very different.

In the Old Testament, dancing is celebrated in the psalms, and it was, of course, normal in wedding feasts and other celebrations, but there was an inevitable conflict with hierarchy and dignity. King David took part near-naked in a dance through the streets of Jerusalem (2 Samuel 6:14), but his wife Michal strongly disapproved. She told him he had dishonoured himself by uncovering himself 'in the eyes of the handmaids of his servants, as one of the vain fellows shamelessly uncovereth himself!' (2 Samuel 6:20). And the Hebrew prophets disapproved of the ecstatic dances that the Jews shared with the other inhabitants of Palestine. Prophet after prophet condemned their dances in sacred groves dedicated to Canaanite goddesses, which could turn into orgies.[40]

There was a similar conflict in ancient Greece between the ecstatic rituals associated with the wine-god Dionysus and the forces of military discipline, highlighted in Euripides' play *The Bacchae*. The warrior king Pentheus tried to crack down on the wildly dancing Maenads, women followers of Dionysus. But in the end, he could not resist; he disguised himself in women's clothes and went to the dance, only to meet a terrible death when he was torn apart limb from limb by his own mother.

Likewise in imperial Rome, wealthy and important people were not supposed to indulge in undignified dancing in public. Nevertheless, the Dionysian cult of Bacchus (the Roman equivalent of Dionysus) became increasingly popular, until it was seen as a threat. It was savagely repressed in 186 BC, when about seven thousand men and women were arrested for taking part in Bacchic rites, and most were executed.[41] The old gods and goddesses were accessible through dancing and ritually induced ecstasy. The newer sky gods, like Yahweh and Zeus, spoke through prophets and priests instead.

With the arrival of Jesus, the situation changed again. Jesus had a Dionysian aspect, and he is commemorated in the Holy Communion through the drinking of wine. His first miracle was the transformation of 180 gallons of water into wine at a wedding feast, after the original supply had already been drunk (John 2: 1-11). Early Christian gatherings often involved feasting, drinking, and probably dancing, too,[42] and a tension between joyful celebration and distrust of disorderly behaviour has run throughout all Christian history.

In the Middle Ages, festivals and carnivals were widely tolerated by the Roman Catholic Church. But at the Protestant Reformation, these popular celebrations were denounced by some of the Reformers, especially by Calvinists. In the seventeenth century, the Puritans in England tried to suppress dancing altogether, and they cut down the maypoles that were the focus for dances in towns and villages. But there were gains as well as losses. Before the Reformation, church congregations played little part in the service, but afterwards, especially in Lutheran Germany, they were encouraged to sing. And Luther himself liked dancing.

Although depression probably existed in the ancient world, Ehrenreich points out that from the seventeenth century onwards, it became an ever more prevalent feature of European culture. Melancholy was on the rise, especially in Protestant countries. A new emphasis on the autonomy of the self gave a greater sense of individual freedom, but it was also isolating. Together with the social disruption experienced by many people as a result of moving from villages into towns and cities, this new individualism was accompanied by increased anxiety and depression.[43]

In the nineteenth and early twentieth centuries, European

explorers and missionaries were often appalled by the ecstatic dances of native peoples, especially when they went into trance, sometimes foaming at the mouth, not feeling pain, seeing visions and believing themselves to be possessed by spirits or deities. In the introduction to a book on tribal dancing published in 1926, the author, W.D. Hambly, had to ask his readers for their sympathy for the subject:

> The student of primitive music and dancing will have to cultivate a habit of broad-minded consideration for the activities of backward races . . . Music and dancing performed wildly by firelight in a tropical forest have not seldom provoked the censure and disgust of European visitors, who have seen only what is grotesque or sensual.[44]

Educated Westerners themselves had more sedate dances, although their less-educated compatriots still danced wildly at Carnival and in other festivities.

In the Caribbean and the Americas in the early nineteenth century, music-making by slaves of African ancestry was not only emotionally disturbing to some white slave owners, it was also politically threatening. Revolts often broke out at times of festival or celebration, including Christmas, when the dancing gave the oppressed people a greater sense of solidarity, community and cooperation. This historical background is one reason why the rock-and-roll revolution that began in the late 1950s had such a profound effect on white society. There was a return of the repressed. The music of white musicians from Elvis Presley onwards was inspired by African-American music, itself deeply rooted in black church music. Black performers of the 1950s and 1960s, such as Ray Charles, Little Richard, and Aretha Franklin, acknowledged their obvious debt to black church music and many of them sang both religious and secular songs.

The rock-and-roll revolution transferred something of an African-American sense of rhythm to people of European descent. And from the 1960s onwards, through music festivals, like the Glastonbury Festival in England, something of the old sense of carnival returned. But apart from the fact that these are secular, rather than religious events, there is a big difference from the older kinds of celebration. People dance in festivals, clubs and at parties, but most do not sing or make music themselves. They are consumers rather than creators. If listening to music alone were enough to counteract depression, then depression would have decreased in recent decades, because music has become all-pervasive through radio, recordings, background music, film soundtracks, commercials, portable music players, and the Internet. But rates of depression have increased rather than diminished.[45]

People who go to church still sing together, as do people who belong to choirs. But the majority of the population in Europe, and now even a majority in the United States, neither sing in churches nor in choirs. This may help to explain the popularity of karaoke, which enables many people to start singing again. Probably any kind of singing is better than none. But singing with a spiritual purpose may be more effective than purely secular singing, because it can lead to a sense of connection not only with other people, but also with more-than-human consciousness, going beyond the human realm to the divine. At least this was the experience of the black gospel singer Mahalia Jackson, who said, 'I sing God's music because it makes me feel free. It gives me hope. With the blues, when you finish, you still have the blues.'[46]

Musical goddesses, gods and spirits

Goddesses, gods, angels, spirits and God are music lovers. Their devotees sing to them, chant to them, invoke them through songs, praise them through psalms and hymns, and the angels themselves are musical beings, as in a well-known Christmas carol, 'Sing choirs of angels, sing in exultation.'

The ancient Greeks thought that goddesses, the muses, inspired the arts. That is why music is called music: it is inspired by muses. In Greek mythology, Orpheus, the legendary musician and archetype of inspired singers, was the son of a muse. Likewise in India, music is believed to be inspired by a goddess, Saraswati, who is usually portrayed playing a *veena,* a stringed instrument. In South India, concerts of classical Indian music usually begin by invoking this goddess.

Jews, Christians and Muslims all believe that God loves music. All three traditions acknowledge the psalms as holy songs, and in the Koran (Surah 4:163) God is identified as their source: 'and to David We gave the psalms'. Many of the psalms are about making music, in some cases by humans and by non-humans, too, as in Psalm 98:5-9: 'Sing unto the Lord with the harp; with the harp, and with the voice of a psalm. With the trumpet and sound of cornet make a joyful noise before the Lord, the King . . . Let the floods clap their hands; let the hills be joyful together before the Lord.'

No fewer than three of the psalms (96, 98 and 149) begin with the words, 'Sing unto the Lord a new song.' The Western church music tradition, both Roman Catholic and Protestant, from the sixteenth century right up until the present day has produced an amazing variety of new songs, some of very great beauty. And God not only likes new songs, but also, as in other religious traditions, old songs, traditional chants and chanted prayers.

Why do spiritual beings like music? Atheists and Secular Humanists have a ready answer. Spiritual beings cannot like music because they do not exist. Humans like music and they then project this human activity onto imaginary gods, goddesses, and angels. In sacred music and chanting, humans are not connecting with higher forms of consciousness, but only with electro-chemical events in their own brains.

By contrast, most if not all religious traditions assume without question that the ultimate reality of the universe is vibratory or sonic and at the same time conscious. In several Hindu accounts, the universe was formed by primal sounds, first and foremost the mantra *Om*. In the Judeo-Christian tradition, God creates by speaking. God's Word, or *logos*, to use the Greek term, is the second person of the Christian Holy Trinity. God the Father is the speaker of the Word. The Holy Spirit is the breath by which he speaks it.

In Plato's 'Myth of Er', which comes at the end of *The Republic*, he describes the soul's journey through the whirling circles of the heavens, which carry the planets, with each planetary level emitting its own note, creating a cosmic harmony.[47] The Roman poet Cicero (106–43 BC) also wrote a book called *The Republic*, partly inspired by Plato, which also included a heavenly journey, called the 'Dream of Scipio', in which Scipio's dead grandfather guided him. He visited the place where departed souls dwell in the Milky Way, enabling him to look back on the planetary spheres as if from outside, rather like a super-astronaut. In his vision of the cosmos, the earth was at the centre, circled by the moon and then the spheres of the other planets, and he heard a 'great and pleasing sound' caused by the motion of the spheres themselves. His grandfather explained, 'Gifted men, imitating this harmony on string instruments and in singing, have gained for themselves a return to this region, as have those of exceptional abilities who have studied divine matters even in earthly life. The ears of mortals are filled with this sound, but they are unable to hear it.'[48]

Of course, we now have a very different cosmology, and the earth is no longer at its centre. The studies of planetary motions by Johannes Kepler (1571–1630) showed that planets move around the sun not in circles but in ellipses. In 1619 he gave an account of the planets' songs that described their real music as polyphonic and not a static scale of notes as in previous visions of the harmony of the spheres. As the planets moved in their elliptical orbits, they speeded up and slowed down, creating an interweaving of tones. Significantly, Kepler published his findings in a book called *Harmonices Mundi* (harmony of the world).

It is, of course, still the case that the planets have elliptical orbits of particular periods or frequencies, and also the sun has an orbital movement within the galaxy, as do the other stars. These frequencies are much too slow to be registered as tones by human hearing, but if there were a galactic mind, then it might well hear the repetitive rhythms of all these celestial movements as tones or qualities, as a kind of planetary, stellar and galactic music.

In the background of all these musical theories of the cosmos were the seminal teachings of the school of Pythagoras in ancient Greece. The Pythagoreans believed that numbers, ratios and proportions underlay the entire cosmos. They also showed that music provided a bridge between quantity and quality, between mathematics – measurable aspects of music – and subjective experience. Musical intervals could be both heard consciously and expressed mathematically. For example, if one flute is twice as long as another, the note it sounds is an octave lower. If it is half as long, the note is an octave higher. The same is true for the length of strings in a stringed instrument (if the thickness and tension are constant). These principles apply to our own vocal cords, too, which are string-like.[49]

Contemporary science follows the same principles, but gives us more detail about the relationship of quantity and quality. If we beat out a rhythm once a second, we hear a series of beats that we can count. But as the beats become faster and faster, by about 20 beats per second (20 Hertz or Hz for short), we can no longer count them but hear low notes instead, qualities rather than quantities. As the frequency increases, the notes get higher and higher. Within a range from about 20 to 20,000 Hz, we hear vibrations as tones, as qualities. Yet they are also measurable quantitatively as frequencies. In the conventional tuning system, the note A above middle C is defined with a frequency of 440 Hz. The note A an octave below has a frequency of 220 Hz; A an octave above, 880 Hz.

Quantum mechanics has extended these Pythagorean principles down to the fundamental particles of matter, which are not made of solid stuff, but are patterns of vibration, as is light. Atoms, molecules and crystals are all vibratory structures. Indeed, everything in nature is rhythmic or vibratory, including our own physiology, with our brain waves, heartbeats, breathing patterns, daily cycles of waking and sleeping, monthly cycles in women, and annual cycles for all of us.

For panpsychists, there may well be many forms of mind or consciousness in nature, each of them experiencing qualities and feelings at its own level. What if wave patterns at many different levels become conscious, from the smallest, in subatomic particles to the largest, in galactic clusters and indeed the entire cosmos? What if quality, namely sounds, and quantity, namely frequencies and amplitudes, go together in minds at all levels of complexity, not just in animal minds? What if all nature can be experienced as music?

The Indian musician and Sufi, Hazrat Inayat Khan (1882–1927) expressed this possibility as follows:

> Music as we know it in our everyday language is only a miniature: that which our intelligence has grasped from that music or harmony of the entire universe that is working behind us. The harmony of the universe is the background of the little picture that we call music. Our sense of music, our attraction to music, shows that music is in the depth of our being. Music is behind the working of the whole universe. Music is not only life's great object, but music is life itself.[50]

Something of the same insight underlies the first part of *The Silmarillion*, by J.R.R. Tolkien, the author of *Lord of the Rings*, which tells the creation myth of the universe Eä that contains

Middle Earth. The story begins with the creation of angel-like beings:

> There was Eru, the One, who in Arda is called Ilúvatar; and he made first the Ainur, the Holy Ones, that were the offspring of his thought, and they were with him before aught else was made. And he spoke to them, propounding to them themes of music; and they sang before him, and he was glad. But for a long while they sang only each alone, or but few together, while the rest hearkened; for each comprehended only that part of the mind of Ilúvatar from which he came, and in the understanding of their brethren they grew but slowly. Yet ever as they listened they came to deeper understanding, and increased in unison and harmony.[51]

Tolkien's poetic imagining of creation through music helps deepen our imaginations. Cosmic music is far beyond our normal range of experience, but creation myths and storytellers help us to glimpse something of a conscious world far beyond our limited minds, yet to which we are related through the shared experience of music.

For people who believe that consciousness exists only inside brains, the appreciation of music must be brain-bound; everything else is unconscious; the vast majority of the non-human world is deaf to our chants, songs and music. On the other hand, if the entire cosmos is conscious, and if it contains many levels of consciousness within it, then music can link us to musical minds far greater than our own, and ultimately to the source of life itself.

Two musical practices

SINGING

Make a practice of singing with other people. The simplest way to do so is to join a community choir, or a church choir, or simply go to church on Sunday. In most churches, you will be able to participate in the singing of hymns and psalms at a morning or evening service. This is what I do myself, wherever I am. It is much simpler than trying to get together a group of friends to sing. If you feel uncomfortable going to a Christian service, then try the Sunday Assembly or some other secular group that meets and sings regularly. If you are Jewish, go to a synagogue with participatory singing. If you are Hindu, go to a bhajan or other singing group.

CHANTING

I asked my wife Jill to summarise some simple practices that all of us can try. Here are her suggestions:

Most spiritual practices are ways of allowing us to be in the present moment, to be here now. We can only chant in the present, and if we listen to the sound we are making as we make it, we create a circuit of attention. This allows us to integrate with the unfolding duration of the now, where joy is to be found. When people talk of being disenchanted, I take this literally and tell them the remedy for disenchantment is to chant. To enchant is to make, and to be made, magical through sound.

All traditions have sacred sounds that are repeated as meditations to rescue us from our exile in the delusion of past and future, from our endless loop of regrets and dreads and bring us back into the now. In the East, there are countless mantras, perhaps the best known is *Om*, while in the West – in Judaism, Christianity and Islam – many have chanted *Ameen*; indeed, some speculate these may have a common origin.

I suggest you try the following practice.

Close your eyes and focus on your breath; let every in-breath bring light into your body, and with every out-breath release tension and let go. With every out-breath let go further and with every in-breath feel inside yourself, until all of you is light. Then continue, but now introduce sound, start to hum and let the sound be like a beam of light, exploring the inner recesses of your being. Most importantly, listen, keeping focused on the sound you are making, so that nothing escapes your hearing. As you continue, begin to change the shape of your mouth, exploring different vowel sounds, moving your tongue around until the sound you are listening to begins to modulate and change.

Choose whichever mantra you feel drawn to and sit quietly while you chant it. Maybe chant the mantra *Ah*, the Tibetan mantra of primordial space, and at the same time listen to yourself, so you integrate with the sound you are chanting. Gradually allow the sound to become quieter till you are present listening to the absence of sound. The gift of sound is silence.

7

Pilgrimages and Holy Places

Thousands of animal species are migratory. They usually have two homes, moving from one to the other in an annual cycle. Swallows arrive in England in the spring, often returning to the very same place they nested the year before. In the autumn, they fly to South Africa. They make the reverse journey the next spring. Their homes are like two poles between which they move. The Arctic tern, a small seabird, literally moves between two poles in its annual migration from the Arctic, where it breeds, to the Antarctic and back again.

These migrations are purposeful. The animals migrate to places with favourable conditions for breeding, and then move to places where they can find food and warmth while it is winter in their breeding grounds.

However, some animals make migratory journeys without any obvious biological purpose. The kingfish of the Mtentu River in West Africa make yearly journeys to the head of the river, where they swim in clockwise circles for a week before returning. They neither breed nor hunt at their destination, and their annual migration has been compared to a pilgrimage.[1] Some groups of chimpanzees carry rocks to particular trees in their territories where they throw them down. The stones accumulate in heaps, rather like human-made cairns.[2]

For most of human history, the vast majority of humanity was migratory. Our ancestors were hunter-gatherers. Hunting and gathering mean moving to find game and edible plants; they involve purposive cycles of movement. Traditional peoples follow customary migratory paths, as the reindeer herders of Siberia still do today.[3] Australian Aborigines navigated these paths, or Song Lines, by singing the story of the places as they travelled, with the songs highlighting the locations of waterholes and landmarks.

In North America, too, hunter-gatherer societies made circuits of their territories to sources of natural resources, and also to places featured in creation songs and stories. Their rituals were linked to specific sacred sites. Paiute-Shoshone people of California believed that a particular hot spring was the site of their creation, and that it was a healing place. The Chumash Indians helped the deceased on their journey by burying medicine bundles on the top of Santa Lucia peak. One of the Sioux legends told how a woman refused to break camp and follow the tribe's migration trail owing to jealousy about her husband's new wife. So she stayed behind, and turned into a standing stone, at a place now called Standing Rock.[4]

From migration to modern pilgrimage

About 12,000 years ago, the Neolithic Revolution began. People started cultivating crops. Since then, an increasing proportion of humanity has led settled lives in villages, and then in towns and cities. For all these people, and for all of us today who live in villages, towns and cities, this immemorial pattern of continual movement has come to an end.

When agriculture and settled life began, the herders of goats, sheep, cattle, yaks and camels continued a migratory existence, moving their herds and flocks in search of water and fresh pastures, going to higher ground in the summer, and to lower ground in the winter. In the biblical account, when Adam and Eve were driven out of the Garden of Eden, one of their sons, Cain, became a farmer, and the other, Abel, a shepherd. Like Abel, the Old Testament patriarchs, Abraham, Isaac and Jacob, were herders, who moved amidst settled people. Humanity at this stage was depicted as half settled and half on a journey.

Before the development of agriculture and settled living, the sacred sites were linked to seasonal festivals, as people moved from one place to another.

The sanctity of local shrines extended to the paths that led up to and between them. For settled people, the ancient habit of making journeys to holy places persisted, and, in some cases, the migratory movement of the group was replaced by ritualised sacred journeys in the form of religious processions.[5]

With the growth of cities, pilgrimages focused increasingly on man-made temples. The cities of the ancient world were sacralised – and justified – by the presence of temples, as in ancient Egypt and Sumeria. In Sumer, all the great city-states had a temple at their centre. Less urban civilisations, as in England, built great ceremonial centres, such as the stone circles of Avebury and Stonehenge, which were constructed more than 4,000 years ago, around the same time as the pyramids in Egypt. These great structures must have been places at which populations converged for seasonal festivals, making journeys that were prototypical pilgrimages.

In his *Republic*, Plato advised settlers in a new country first to discover the shrines and sacred places of the local deities, and then reconsecrate them to the corresponding principles in the settlers' religion, with festivals on the appropriate days.[6] By Plato's time, many religions had already adopted this principle, and many did so subsequently, including the Orthodox and Catholic Churches.

Moses and Joshua led the Jewish people out of slavery in Egypt to the Promised Land – namely Canaan or Palestine. When they settled there, they originally paid homage at a range of holy places that had been venerated long before their arrival, like Shiloh, a Bronze Age shrine sacred to the Canaanites, where Joshua set up the holy tent. The Jewish people also worshipped in sacred groves on hilltops, and venerated the sacred stone at Bethel where Jacob had his vision of angels descending from and ascending into heaven. Many other megaliths in Palestine were sacred to the pre-Jewish inhabitants of the land. Bethel may well have been one of these ancient sacred stones when Jacob had his vision there.[7]

Jacob had anointed this stone with oil and established an altar there. Later, his descendants became slaves in Egypt. When they returned to Canaan after many generations, they made Bethel a major place of pilgrimage.

After the construction of the temple in Jerusalem by King Solomon, around 950 BC,[8] Jerusalem became a central place of pilgrimage, especially at the time of the great festivals. More than 200 years later, King Hezekiah, who reigned from 715–687 BC, destroyed the hilltop shrines and other sacred places, and tried to channel all pilgrimage to the temple at Jerusalem. But he failed to suppress worship at Bethel, which continued to rival Jerusalem as a religious centre until the reign of King Josiah (640–609 BC), who completed the centralisation of Jewish worship by destroying the sanctuary at Bethel and cutting down the remaining sacred groves. The focus of Jewish pilgrimage was henceforth urban, at the temple in the city, rather than spread out over many groves, shrines and other sacred places. But the principle of pilgrimage remained.

In classical Greece, each city-state had its central temple to which far-flung citizens returned for regular festivals. In Athens, the Great Panathenaia festival, celebrated every four years, culminated in a procession on the Acropolis, which is represented on the friezes of the Parthenon, the temple of Athena, the patron goddess of Athens.[9] And as well as these local gatherings, there were all-Greek centres of pilgrimage, like the shrine of Delphi, where pilgrims consulted the oracle, and Olympia, where the Olympic Games took place every four years at the festival of Zeus. Here the people would see their champions perform feats of strength, speed and endurance, embodying heroic myths in flesh and spectacle.

The classical Greek traditions also included another core purpose of pilgrimage: personal healing. Many pilgrims went to the great healing shrine at Epidaurus, hoping for miraculous healings by the gods Apollo and Asclepius. Pilgrims made offerings and slept inside the shrine, where many claimed to have been cured in visions.[10]

This tradition of dream incubation continued within the Greek Orthodox Church, particularly in churches dedicated to the twin healing saints Cosmas and Damian, who were said to have performed the first successful full-leg transplant to an amputated limb. This tradition continues in some Orthodox churches and monasteries, where pilgrims sleep the night in the hope of receiving divinely inspired dreams and cures.

Similarly, to this day, Muslim pilgrims sleep at the shrines of Sufi saints in the hope of receiving healing dreams. When I was living in Hyderabad, India, some Muslim friends took me to the shrine of a local saint, within an old caravanserai, a walled compound where travellers could rest. The shrine was in a courtyard, and family groups were scattered around preparing to stay the night, to support a troubled member of the family. They were praying for healing dreams, in which the saint would appear to them and help them. I was told that many people received healing dreams.

Within the Roman Empire there were many places of pilgrimage. Some were by springs, rivers and sacred groves that only local people would visit; others were more widely famous, and involved many days' journey on foot. Some pilgrims' practices were similar to those of present-day Buddhist monks. For example, a treatise from the second century AD called *On the Syrian Goddess* describes how pilgrims prepared themselves for their journey to the holy city of Hierapolis, in modern-day Turkey, by shaving their heads and eyebrows before setting off. On their way, they always slept on the ground, never on a bed, and used only cold water for bathing.[11]

For early Christians, the primary place of pilgrimage was Jerusalem, because of its central importance in Jesus's life, death and resurrection. Jesus himself travelled around the Holy Land on foot, and went to Jerusalem for the major festivals.

One of the first and most important Christian pilgrims was the Empress Helena (AD c.250–c.320), who went to find the significant places in the life of Jesus, and to look for relics, including the cross on which Jesus was crucified. Her son Constantine converted to Christianity, founded Constantinople as the capital of the Eastern Roman Empire, and built the Church of the Holy Sepulchre on the site where Jesus was believed to have been buried and raised from the dead.

Jerusalem is still a primary place of pilgrimage for Christians and for Jews, and for Muslims. In his visionary 'night journey', Mohammed flew to the temple mount in Jerusalem on a steed called Lightning, where, according to Muslim tradition, he encountered Abraham, Moses, Jesus and other prophets. He led them in prayers. Gabriel then escorted Mohammed to the pinnacle of the rock where a ladder of golden light appeared. On this glittering shaft, Mohammed ascended through the seven heavens into the presence of Allah, from whom he received instructions for himself and his followers. Over this place stands the Dome of the Rock, one of the holiest places in Islam.

In the Christian world, many additional places for pilgrimage grew up around the tombs of martyrs and other saints, whose relics were believed to connect the pilgrim to the heavenly realm to which the saints had ascended. Their tombs were seen as places where heaven and earth joined. Through their earthly relics, the saints in heaven could be present at their tomb on earth. These tombs were already places of pilgrimage by the third century AD, and by the sixth century the graves of the saints had become centres of ecclesiastical life. In the Western Church, the power and authority of bishops was closely linked to the shrines of saints, which were often housed in cathedrals.[12]

The hometown of the prophet Mohammed, Mecca, was already an important place of pilgrimage at the time of his birth, with the pilgrimage centred on a black rock, which tradition held had fallen from heaven. This black stone is now embedded in one corner of the Kaaba, the cubic building at the centre of Mecca, the focus of Islamic pilgrimage, around which pilgrims walk seven times anti-clockwise. This is one of the few places in the world where circumambulation does not go clockwise.

India is still criss-crossed with numerous pilgrim routes leading to holy caves, like Amarnath high in the mountains of Kashmir, to the sources of sacred rivers such as the Ganges, to holy mountains, like Mount Kailash in Tibet, and to many temples, sacred trees, rivers, rocks and hilltop shrines. Buddhists go on pilgrimages to Buddhist sacred places, including those linked to the Buddha's life in India, like Bodh Gaya, the place where the Buddha is said to have attained Enlightenment, under a *bodhi* or *pipal* tree.[13]

Forms of pilgrimage are found all over the world. Pilgrimage seems to be a deeply ingrained part of human nature, with its roots in the seasonal migrations of hunter-gatherers, and, more remotely, in many millions of years of animal migrations.

Precisely because of its ancient roots, pilgrimage was attacked and suppressed in Europe in the Protestant Reformation. The reformers based their faith on the authority of the Bible, rather than on the pre-Christian traditions that had over the centuries been syncretised and absorbed within the Catholic Church, together with newer customs and traditions related to Christian saints.

In England, there were great pilgrimages to the shrine of St Thomas Becket in Canterbury Cathedral, commemorating his martyrdom, and all it symbolised of spiritual resistance to earthly and especially royal power. Thomas was also known as the Great Doctor, a healer without compare in days without affordable doctors or medical science. The healing power was supposed to reside in water tinged with his blood, 'the blood of St Thomas', which pilgrims bought in lead ampullae from vendors near his shrine. The journey to Canterbury was immortalised in Chaucer's *Canterbury Tales*, written in the 1380s and 1390s, consisting of stories told by pilgrims to each other on their journey.

Another great English centre of pilgrimage was Walsingham, in Norfolk, where there was a shrine of the Blessed Virgin Mary in the form of a Black Madonna, and her Holy House, a recreation of the building in which the Annunciation by the Angel Gabriel happened. Another was the great Abbey of Glastonbury, where King Arthur was reputedly buried and where Joseph of Arimathea (who arranged the burial of Jesus after his crucifixion), is said to have planted his staff in a nearby hill, where it took root and grew into the Holy Thorn Tree that flowered on Christmas Day. Regular hawthorns flower in May. Trees grown from cuttings said to be descended from the original Holy Thorn still flower in Glastonbury at Christmas.

But there was nothing about Canterbury, Walsingham or Glastonbury in the Bible, and therefore for the Protestant Reformers these pilgrimages were invalid. They had no scriptural authority.

In 1538, all English pilgrimages were suppressed under King Henry VIII by his henchman Thomas Cromwell. The injunction against pilgrimage expressed an austere Protestant spirit:

[The people should] not repose their trust and affiance in any other works devised by men's phantasies beside Scripture; as in wandering to pilgrimages, offering of money, candles or tapers to images or relics, or kissing or licking the same, saying over a number of beads, not understood or minded on.

The shrines were destroyed, the abbeys and monasteries dissolved, and their riches confiscated by the king. The dissolution of the monasteries doubly destroyed the pilgrimage landscape by removing key pilgrim destinations, and also by taking away the infrastructure that supported pilgrims as they travelled, providing them with food and accommodation.

Pilgrimages were also suppressed in other Protestant countries. In 1520, Martin Luther declared, 'All pilgrimages should be stopped. There is no good in them: no commandment enjoins them, no obedience attaches to them. Rather do these pilgrimages give countless occasions to commit sin and to despise God's commandments.'[14]

No such suppression occurred in the Roman Catholic countries of Europe, or in the Orthodox East. In many Catholic and Orthodox countries, ancient pilgrimages continue to this day. In Ireland, despite attempts to suppress them by the Protestant English, pilgrimages have persisted, and many pilgrims still go to the island Sanctuary of St Patrick in Lough Derg, in County Donegal, and climb the holy mountain, Croagh Patrick, in County Mayo.

The most famous European pilgrimage is to Santiago de Compostela, in Spain. It has not only persisted from the Middle Ages, but has undergone a huge revival in the last thirty years, as discussed below.

In Latin America, the European conquerors followed the traditional Roman Catholic policy of assimilating and Christianising pre-Christian holy places. Near Mexico City, the temple of the Aztec mother goddess was demolished in 1519. Then in 1531, a native Mexican had a series of visions of the Blessed Virgin on the same spot, where a shrine was built, which is now the Basilica of Our Lady of Guadalupe, a Black Madonna standing on a crescent moon. This is the most visited Roman Catholic place of pilgrimage in the world.

By contrast, the Protestant settlers of North America were not interested in the holy places of the native peoples. English common law was taken to redefine the native people's homeland as *vacuum domicilium*, an unpopulated expanse of wilderness over which no one held dominion, which cried out for farming and civilisation.[15]

In some traditionally Roman Catholic and Orthodox countries, pilgrimages were suppressed not by Christian reformers, but by anti-Christian revolutionaries. They wanted to stamp out pilgrimage precisely because it was religious. The French Revolution, starting in 1789, aimed to overthrow the power of the Roman Catholic Church as well as the power of the king. In 1793, the revolutionary government proclaimed the Cult of Reason as the state religion. The Cathedral of Notre Dame in Paris was converted into a Temple of Reason; other churches and cathedrals were secularised. Pilgrimage was banned.

Under the atheist government of the Soviet Union, churches were closed, priests executed, monasteries destroyed and religious activity persecuted. Shrines were deliberately desecrated, most recently in the campaign against 'so-called holy places' launched in 1958, with the aim of the final elimination of pilgrimage.[16]

Yet the communists did not dismiss the idea of pilgrimage to relics; they had their own version. Some Soviet visionaries were convinced that science would overcome physical death, and confer immortality on humans, enabling them to live forever. When Lenin died in 1924, an official Immortalisation Commission was set up to investigate how his body could be preserved until he could be brought back to life. He was embalmed in the hope that he would be preserved long enough to be resurrected, just as some American millionaires have their whole bodies (or, cheaper, only their heads) cryogenically frozen in the hope of resurrection.[17]

Lenin's body was placed in a mausoleum in Red Square in Moscow, which rapidly became a centre of communist pilgrimage.

Millions of people visited Lenin's tomb during the Soviet period, and it still attracts visitors today. They are officially required to show respect: men have to remove their hats; and talking and photography are forbidden.[18] Likewise in Beijing, there is a mausoleum to Mao Zedong in the middle of Tiananmen Square that also attracts a continual stream of pilgrims, who file past Mao's embalmed body and make offerings of flowers.

Objections to pilgrimage

Although Protestants and political revolutionaries tried to suppress religious pilgrimages, they were not the first to find fault with them. Over the centuries there were four main objections to pilgrimage by religious people themselves.

1. The first and most profound objection was that pilgrimage is unnecessary. God is everywhere, and human beings can pray to God wherever they are. There is no need to go to special places. In the fourth century AD, St Gregory of Nyssa put it as follows: 'Change of place does not effect any drawing nearer to God, but wherever you may be God will come to you.' His contemporary St Jerome agreed: 'Access to the court of heaven is as easy from Britain as it is from Jerusalem.'[19] So there was no need to travel.

    Some people opposed physical pilgrimage by internalising it, so that the whole of a person's life was seen as a pilgrimage. The most famous example of this approach in a Protestant context is *The Pilgrim's Progress* by John Bunyan, a seventeenth-century English Baptist preacher. However, the idea of life as a pilgrimage is a metaphor that depends on actual pilgrimage. For those who have never been on a pilgrimage, the metaphor is a mere idea, no longer grounded in a lived experience.

    Some people may have passed beyond the need for pilgrimage,

because they have found a way of living in the presence of God wherever they are. But perhaps some of them reached this state through going on pilgrimages in the first place. This argument against pilgrimage is about passing beyond it, rather than not starting it.

2. Pilgrimage is idolatrous. The second of the Ten Commandments, in the words of the King James Bible, reads: 'Thou shalt not make unto thee any graven image.' If pilgrims were going to worship man-made idols, then pilgrimage would be idolatrous. But relics of saints were not graven images, nor were holy wells, nor sacred stones.

But what about icons? In the early church, icons were widely used as aids to prayer and meditation through connecting with a visual image of Christ or of the saints. An important argument in defence of icons was that Jesus Christ was the incarnation of God. God had taken a human form, and therefore the representation of the human form was not idolatry.

In AD 730, the Byzantine Emperor Leo III prohibited the use of icons. The empire was traumatised by the ensuing outburst of iconoclasm, which literally means the destruction of images (from the Greek *eikon* = image and *klastes* = breaker). But in 787, the Emperor Cyrene re-established the use of icons. After another burst of iconoclasm from 815 to 843, the Empress Theodora reinstated them again. The Eastern Orthodox Church celebrates this final restoration of icons on the Feast of Orthodoxy, on the first Sunday of Lent.

Iconoclasm re-emerged in the Protestant Reformation. In England, many images of saints and angels as well as stained-glass windows were destroyed in the reign of King Henry VIII, under the administration of Thomas Cromwell. There was a second wave of iconoclasm under the rule of his Puritan name-sake (and relative) Oliver Cromwell, when England became a republic, rather than a monarchy (1649–60). But many paintings, crucifixes, statues and stained-glass windows survived. And after the Restoration of the Monarchy in 1660, religious images were reinstated in the Church of England.

Iconoclasm emerged yet again in the twenty-first century in the destruction of giant Buddhist statues in Afghanistan by Taliban militias and of ancient artefacts in Iraq and Syria by the Islamic State.

But the question of idolatry is irrelevant to many forms of pilgrimage, including those to sacred groves, holy wells and mountains, sacred stones, and relics of saints. Many of the foci of pilgrimage are not images made by human hands.

3. Pilgrimage is superstitious. It is primitive, outmoded, ignorant, and has been superseded by a higher level of understanding or enlightenment.

Superstition literally means standing over, or survival.[20] From the point of view of the early Christians, the practices of other religions were superstitious. From the point of view of the Protestant reformers, the practices of the Catholic Church were superstitious. From the point of view of Enlightenment intellectuals, all religious practices were superstitious, and were suppressed as part of the French Revolution in the Reign of Terror. Similarly, the atheist governments of the Soviet Union, Communist China, and Cambodia under Pol Pot, treated all religious practices as superstitious, and suppressed them in favour of the Marxist philosophy of materialism.

The dismissal of pilgrimage as superstitious is a consequence of anti-traditional or anti-religious worldviews and says more about these worldviews than about pilgrimage itself.

4. Pilgrimages are occasions for adultery, fornication, drinking, commercialisation and other disreputable activities. Chaucer's

Canterbury Tales abound with stories of sexual licence, and no doubt this was not merely a matter of storytelling.

This particular objection to pilgrimage seems irrelevant in the Western world today, though it may still have some validity elsewhere. In modern secular societies, no one needs to go on a pilgrimage to engage in sexual adventures, and secular tourism leads to even more commercialisation and hustling than pilgrimage.

Pilgrimage and tourism

Although pilgrimage was suppressed in Protestant countries and by revolutionary governments, the urge to visit holy places was not extinguished. Within two hundred years of the banning of pilgrimage in England, the English had invented tourism, now a vast global industry, with a global economic value of 2.2 trillion dollars in 2013.[21]

Tourism is often a form of frustrated pilgrimage. Many tourists still go to the ancient sacred places: pyramids and temples in Egypt, Stonehenge, the great cathedrals of Europe, the temples, sacred rivers and mountains of India, ancient sacred places like Uluru (formerly known as Ayers Rock) in Australia, the Temple of the Sun in Machu Picchu, Peru, and so on. But whereas pilgrims traditionally walk to their destinations, often enduring hardships on the way, tourists travel in cars, buses and planes. They are not going to sacred sites to make offerings, or pray. Many feel they have to behave as secular, modern people who are primarily interested in cultural history. Guides deluge them with historical details that go in one ear and out the other.

What are the key differences between travelling as a tourist and travelling as a pilgrim? Both go to the same kinds of places, but with different intentions. Pilgrims go to connect with a holy place; reaching that holy place is the purpose of their journey. They go with an intention to give thanks for some blessing they have received, or to pray for some blessing they want to receive, or as an act of penance to make amends for something they have done wrong, or for healing, or for inspiration. The tourist goes to see a new place, to hear something of its history, and to take photographs. But they are still making purposive journeys and the ancient holy places still pull them; indeed, these places are often called tourist 'attractions'.

Often when pilgrims return home, they bring something back with them to share with others, including them in the blessings they have received. In India, many pilgrims return from pilgrimages with *prasad*, holy food offered to a god or goddess and blessed in a temple, which they share with their families and friends. Medieval pilgrims in Europe also brought home objects from the sacred centres to which they journeyed, often in the form of badges. These would perform a double function, being focal points of the blessing received, and also visual markers of their wearer's prestige in having made a difficult sacred journey. Tourists likewise return with souvenirs and photographs, but they cannot pass on blessings they have not received themselves.

Although we live in an unprecedented era of mass tourism, in recent decades there has been a remarkable revival of pilgrimage.

The revival of pilgrimage

Even those early Christians who condemned local pilgrimages found it hard to condemn pilgrimages to the Holy Land. St Gregory of Nyssa, despite the many corruptions surrounding them, called these holy places 'memorials of the immense love of the Lord for us men.' And it was through pilgrimages to Jerusalem in the nineteenth century that pilgrimage once again became respectable in the Protestant world. A pilgrimage to the Holy Land in 1862 by Prince Edward, son of Queen Victoria, later King Edward VII, gave it a royal seal of approval and respectability. In 1869, Thomas Cook started organising pilgrimage groups to the Holy Land, and this was the origin of the global travel agency that bears his name. The original package holiday was a pilgrimage. Soon afterwards, Prince George, later King George V, went on a pilgrimage organised by Cook. In England in the late nineteenth century, local pilgrimages to cathedrals began again, and in the twentieth century several places of pilgrimage that were suppressed at the Reformation were revived, like the shrine of Our Lady of Walsingham in Norfolk.

In the Middle Ages, Chartres Cathedral, about sixty miles from Paris, was one of the most important places of pilgrimage in Europe. Even before the cathedral was built, Chartres was the focus of a pilgrimage to a sacred well, the 'Well of the Strong Saints'. The cathedral was built around it, and the well can still be visited in the crypt. Nearby, also in the crypt, is a shrine of Our Lady of Chartres, a Black Madonna. At the time of the French Revolution, pilgrimage stopped abruptly. The modern revival of pilgrimage began only in 1912 when the poet Charles Péguy went to Chartres on a pilgrimage with a group of friends, and wrote a book about it that became a bestseller. Now tens of thousands of pilgrims a year go to Chartres, some on a three-day walking pilgrimage from the Cathedral of Notre Dame in Paris.[22]

Santiago de Compostela, where the cathedral houses the supposed relics of St James, the patron saint of Spain, was one of the most popular sites of pilgrimage in medieval Europe. Numbers are hard to estimate, but some indication of the scale is given by the fact that the monastery of Roncesvalles, one of the first resting places in Spain for French pilgrims who had crossed the Pyrenees, fed some 100,000 pilgrims a year.[23]

The numbers of pilgrims from Northern Europe travelling to Santiago de Compostela plummeted following the Protestant Reformation. Moreover, in the subsequent war with Elizabethan

England, Sir Francis Drake led a naval raid on the nearby city of La Coruña in 1589, and the Archbishop of Santiago hid the relics of St James so that the English could not capture them. He hid them so well that the shrine was empty for nearly three hundred years, and the number of pilgrims was reduced to a trickle. The relics were not rediscovered until 1879, and after being authenticated by Pope Leo XIII, were replaced under the High Altar in 1884, where they remain to this day.[24] However, the pilgrimage itself was not revived until 1949, when a small group of French scholars went on a pilgrimage that was filmed and shown on television in the 1950s, helping to reawaken interest. Even so, the numbers of pilgrims were small. In the 1980s, a few enthusiasts made sure that the way, or Camino, was well marked with signs, and established a series of facilities for pilgrims along the route.

What happened next was remarkable. Here are the annual numbers of pilgrims as recorded by the Spanish authorities:

Number of pilgrims travelling to Santiago de Compostela on foot, by bicycle or on horseback

Year	Number
1987	1,000
1991	10,000
1993	100,000
2004	180,000
2015	263,000

The great majority of these pilgrims walked, but a minority, around ten per cent, travelled by bicycle and a few, less than one per cent, went on horseback.[25] These numbers do not include people who travelled to Santiago by plane, train, bus or car.

As in the Middle Ages, there is now a wide network of pilgrimage routes to Santiago, from several starting points in France, including

Vezelay, in Burgundy, and Paris; from Portugal; and from several places in Spain itself. All twelve of the major waymarked routes are called 'Camino de Santiago'.

Elsewhere in Europe, old pilgrimage routes are being re-opened. In Norway, the most important medieval pilgrimage centre was the shrine of St Olaf in Trondheim, to which pilgrimages began soon after his death in 1030, and in the Middle Ages became immensely popular. But when the Lutheran reformation reached Norway in 1537, pilgrimages were banned, and they ceased until the late twentieth century, when increasing numbers of people began walking to Trondheim, once again as pilgrims. In the 1990s the path from Oslo to Trondheim, about 400 miles long, was signposted, and it was officially open by Crown Prince Haakon in 1997.[26]

Meanwhile in Wales and Scotland, old pilgrim ways are being re-established. In England, the British Pilgrimage Trust, of which I am a patron, is helping to re-open the ancient pilgrimage footpaths, first and foremost the pilgrim's way from Southampton to Canterbury over the South Downs, a 220-mile (350-km) walk, taking 18 days, connecting 65 churches, three cathedrals, 75 prehistoric sites, five holy wells, 15 ruined priories, monasteries or abbeys, eight rivers, ten holy hills, five castles, 50 villages, 40 pubs, eight towns and four cities.[27]

In Russia too, after the end of communist rule in 1991, many Russian Orthodox churches and monasteries re-opened, and ever-increasing numbers of pilgrims make their way to the holy places.

I have been on several pilgrimages in India, in continental Europe, in Britain and to the Holy Land. Like many other people who go on such journeys, I have found them inspiring, and they have often been times of great happiness. Some of my recent pilgrimages have been with a godson. I have stopped giving birthday and Christmas presents, because most people have too many possessions. Instead I give experiences. When my godson was 14, I suggested that his present could be to go on a pilgrimage with me to Canterbury, along the last 10 miles of the old pilgrim path, starting at a village called Chartham. I did not know whether this would appeal to him or not, but he accepted enthusiastically.

We took the train to the small station at Chartham, and set off through fields, woods, orchards and meadows. We had a picnic lunch on Bigbury Hill, an Iron Age hill fort, and passed through the village of Harbledown, where we looked for the Black Prince's Well in the grounds of a medieval almshouse. It was so overgrown we were only able to find it with the help of an old lady who lived in one of the almshouses. It was a small spring in a recess under an old stone arch, with mossy steps going down to it.

By the time we reached Canterbury my godson was flagging; he was not used to walking so far, and we had a short rest. We then walked round the cathedral clockwise, circumambulating it, making it our centre. We then entered the cathedral and lit candles at the place of St Thomas Becket's martyrdom, and in the dark, mysterious crypt. We prayed, then went to a tearoom for tea. We returned to the cathedral for choral evensong, which was extremely beautiful. Then we went home to London by train. It was an extraordinarily happy day for both of us.

This set a precedent, and in the following years we followed the same pattern, walking to a cathedral, walking around it clockwise, visiting the shrine and having tea before attending choral evensong. When he was 15, we went to Ely Cathedral, in Cambridgeshire, walking along the banks of the River Cam as this great medieval building loomed large over the flat fenland countryside. We lit candles and prayed at the shrine of St Etheldreda, a seventh-century Anglo Saxon female saint, which was a major place of pilgrimage in pre-Reformation England. In 2016, when my godson was 16, we walked to Lincoln Cathedral about 8 miles along a footpath at the top of the Lincoln Edge, an escarpment of Jurassic limestone, overlooking the valley of the River Trent. The final approach to the cathedral was up the cobbled Steep Hill, the medieval pilgrim street. Finally, we entered the great sacred space, prayed and lit candles at the shrine of St Hugh, and went to choral evensong.

The contemporary reawakening of pilgrimage in Europe is remarkable. As societies become increasingly secular and materialistic, this ancient spiritual practice is undergoing an astonishing revival.

Benefits of pilgrimage

There have been few specific scientific studies of pilgrimage, but the evidence so far suggests that pilgrimages have a beneficial effect in reducing anxiety and depression.[28] But there are countless personal stories about inspirations and healings. Also many pilgrims find that as they travel on foot, they meet other pilgrims, and non-pilgrims, in a socially levelled way. Normal distinctions of wealth, education and social class seem less relevant. And local pilgrimages have the great advantage of being low-cost and accessible to everyone capable of walking.

Most scientific studies that relate to pilgrimage are generic. Indeed, some of them merely prove the obvious. But it is reassuring to know that the obvious is also scientifically observable.

First, walking itself has many proven benefits. It promotes mental health and wellbeing, improves self-esteem, mood and quality of sleep, and reduces stress, anxiety and fatigue.[29]

Second, people who take exercise in the fresh air and in green spaces tend to benefit more than those who exercise indoors, as discussed in Chapter Three.[30]

Third, purposeful activity is more satisfying and contributes more to wellbeing than purposeless activity; this is a basic tenet of the practice of Occupational Therapy.[31]

Fourth, physical exercise protects against depression and other kinds of ill health.[32]

Fifth, healing is influenced by people's hopes and expectations. The placebo effect is very powerful, and shows up strongly in drug trials, particularly if patients and doctors believe that they might be taking a new wonder drug.[33] If pilgrimages increase people's hopes and expectations, which they do, we might expect that visits to holy places would result in numerous healings, which they do. When supportive people surround those who hope for healing, sharing their hopes and expectations, the effects are stronger still.

Over the years, the Roman Catholic Church has emphasised the role of the saints in healing. Canonisation to sainthood itself requires at least two posthumous miracles, often involving physical healing, and millions seeking hope and healing have made pilgrimages to holy places such as Lourdes, in France.

Lourdes, in the foothills of the Pyrenees, rose to prominence in 1858 as a result of apparitions of the Blessed Virgin Mary to a peasant girl at a grotto. The virgin told her to dig in the ground, from which a spring of water began to bubble up. Cures began almost immediately as people drank the water. Today the flow is much greater, and pilgrims bathe in the thousands of gallons that gush from the ground. Lourdes is one of the most important places of pilgrimage in Europe, with about six million pilgrims a year.[34]

Many thousands of people claim to have been cured there miraculously. The official Lourdes Medical Bureau investigates claims of cures in a rigorously scientific spirit, and some of the claims have been very well authenticated.[35] Even sceptics admit that some very sick people get better at Lourdes, although they do not call these healings miraculous; they think of them as examples of the placebo effect, or as 'spontaneous remissions'.

If a pilgrimage helps someone to get better, then the pilgrim's prayers have been answered. Calling the cure a spontaneous remission leaves the remission unexplained. If faith in God and in the

Holy Mother of God makes spontaneous remissions more probable, then faith works.

Less easy to document is the inspiration and encouragement that many people receive from going on pilgrimage. To travel with the purpose of being in a holy place, and then to be in that place can have transformative effects, and give the experience of spiritual connection. Why?

What makes holy places holy?

Holiness is about connection and relationship. The word comes from a root that means whole or healthy. We are not holy when we are separated and disconnected from each other, from the more-than-human world and from the source of all being. We experience that which is holy when we are connected to the source of life that goes far beyond our own limited natures. Some places evoke this experience more than others, whether because of their physical nature, or because of their human associations, or both.

Some places are holy because they are naturally numinous, like some mountaintops, or springs, or waterfalls, or caves. For example, Glastonbury Tor is a striking hill that rises up above the low-lying land around it. It would stand out and attract the eye even if did not have a medieval tower at its summit. Uluru, or Ayers Rock, is a large sandstone structure, an 'island mountain', surrounded by relatively flat land in central Australia, and appears to change colour throughout the day. Glowing red at dawn and sunset, it is an obvious and very striking landmark, of great cultural importance for the indigenous people of that area, and now a major tourist attraction.

Some places may have a particular power because of their orientation, or because of underground water flows, or underground flows of electricity, called telluric currents, or because of their connection with the surrounding landscape. The properties of these places depend on their connections with their terrestrial surroundings, and also on their relationship to the sky and the heavenly bodies.

In some cultures, specialist diviners evaluate the powers of places, and in some cases help to decide where temples, shrines or tombs are constructed. In Europe this art is called *geomancy;* in China *feng-shui,* which literally means 'wind and water'. The techniques of geomancy are not easily translated into conventional scientific terms, but include an understanding of the relationships of the topology and the flows of energy through the landscape. Joseph Needham summarised some of the principles of traditional *feng-shui* in his *Science and Civilisation in China*:

> The forms of hills and the directions of watercourses, being the outcome of the moulding influences of winds and waters, were the most important, but, in addition, the heights and forms of buildings and the directions of roads were potent factors. The force and nature of the invisible currents would be from hour to hour modified by the positions of the heavenly bodies, so that their aspects as seen from the locality in question had to be considered.[36]

Many holy places are a bridge between heaven and earth; they connect the earth to the sky. They are a kind of gateway, as in Jacob's dream at Bethel (Genesis 28:10–19). Standing stones played this connecting role in megalithic cultures, and in ancient Egypt standing stones took on a particularly refined appearance in the form of obelisks, tapering columns with a pyramidal top, often made of a single stone. In setting up standing stones or obelisks, or in building towers and spires and minarets, humans create places that have a literal vertical dimension.

In the ancient sacred groves of the Holy Land, there were sacred trees or poles sacred to the mother goddess Asherah, which were major sites of Jewish worship until they were condemned by the prophets and destroyed in the reigns of kings Hezekiah and Josiah. In many Hindu temples, there are often metal-clad flagpoles in front of the main shrine, called *dwajasthambam,* which are said to connect the heavens to the earth. Many Christian churches have towers or spires, and many mosques are accompanied by minarets.

Symbolically, all these structures link the heavens and the earth. But the connection is more than symbolic: it is literal. Precisely because these structures reach up towards the sky, they attract lightning. They have always acted as channels for very real energy coming from the sky into the earth, and from the earth into the sky. Nowadays they have lightning conductors attached to them for this very reason. Electricity is polar. The movement of electric charge is a two-way process. As the negatively charged pathways of ionised air – called step leaders – move downwards from clouds towards the earth, the strong electric field induces tall objects to send out positively charged 'streamers' that grow towards the cloud. They often have a purple glow. But not all positive streamers make contact with a step leader. They wait. Then step leaders close the gap with some of them and lightning strikes.

Lightning also strikes tall, natural structures such as mountain-tops, and walkers are advised to stay away from summits and pinnacles during thunderstorms.[37] Trees are often channels for lightning, and some species, including oak and ash, are more often struck than others, such as birch and beech. One reason for the sacredness of oaks in Druidic times may well have been their proneness to lightning, and they were sacred to the thunder god both in Northern Europe – to Thor in Scandinavia – and in ancient Greece, to Zeus. A place where lightning has struck acquires a special quality in the eyes of many different cultures. Significantly, an intriguing book about Native American sacred places in the United States is called *Where the Lightning Strikes.*

Until about two hundred years ago, most lightning-attracting structures were religious buildings, like church spires or minarets. In the nineteenth century, large secular structures were erected, such as the Washington Monument in Washington DC, the world's largest obelisk at 554 feet high, and the Eiffel Tower in Paris, 1,063 feet, and these too are major attractors of lightning. In the twentieth century, the tallest buildings were skyscrapers, which are now the principal magnets for lightning in cities. But in many smaller places, religious buildings remain the main attractors. In my home-town, Newark-on-Trent, the spire of the parish church of St Mary Magdalene is 236 feet high. It was completed around 1350, and is still by far the tallest structure in Newark, continually channelling lightning into the ground beneath this holy place.

Until recently, the scientific explanation for lightning concentrated on the electrical potential difference between thunderclouds and the ground, treating it as a local phenomenon. However, the ancient intuition that lightning links the heavens and the earth turns out to be correct. The electric charge on the clouds is linked to electrically charged regions some fifty miles higher in the sky. Electric discharges called sprites, glowing orange or red, pass between thunderclouds and the upper atmosphere. The upper atmosphere itself is greatly influenced by the solar wind, a stream of charged particles released from the sun, and the speed and density of the solar wind depends on solar activity such as flares.

This 'space weather' affects the Northern and Southern Lights, which are themselves plasma discharges, and also influences the amount of lightning on earth: the stronger the solar wind, the greater the number of lightning discharges.[38] Lightning discharges are also increased by influences from much further away, particularly the cosmic rays from supernovas, or exploding stars. Thus lightning literally comes from the heavens, and is channelled through tall structures into the earth. The places where it strikes are literally charged.

All tall buildings are struck by lightning, although in very few cases do people record when this happens. But this is now technically possible. Lightning strike recorders are commercially available, and they detect when a surge of current goes down a lightning conductor. Some even send an SMS message when a strike occurs. If I were running a church, temple or minaret, I would install one of these devices and make the data available online. There are already fascinating online maps and archives of lightning strikes in many parts of the world, together with real-time updates,[39] but they give broad brushstrokes and do not focus on specific places.

In building temples, cathedrals, churches and mosques, people make structures that are explicitly related to God, or ultimate being, or the source of all health and holiness. And shrines that commemorate holy events, holy people and holy deeds give a link to the source of their holiness. In many cases this link is provided by physical relics, such as the Buddha's tooth in the Temple of the Tooth in Kandy, Sri Lanka, or the relics of saints, usually bones, in many cathedrals and churches. The traditional idea that these bones provide a direct link with the life of the person whose bones they are, has taken on a new lease of life with DNA analysis. Even very ancient bones, like those of Neanderthals from 400,000 years ago, contain DNA that can be analysed using modern molecular techniques.

A skeleton discovered in Leicester in 2012 had several signs of being the remains of King Richard III of England, who died in 1485, and DNA recovered from the bones enabled his identity to be confirmed with a high degree of probability. He was reinterred in Leicester Cathedral in 2015. Richard was a king rather than a saint, but no doubt many venerated relics of saints contain traces of the saints' DNA.

Ironically, relics of extinct species in the form of bones and skeletons play a central role in cathedrals of science, like the

Natural History Museum in London, which are like centres of scientific pilgrimage.

Finally, holy places may be holy because they contain a kind of memory of what has happened there before. If many people have prayed, or experienced healing, or been inspired in a holy place, this makes it more likely that others will be affected positively by this place. According to the hypothesis of morphic resonance (discussed in Chapter Five), people in a particular state of sensory stimulation resonate with those who have been in a similar state before. When we enter a holy place, we are exposed to the same stimuli as those who have been there before, and therefore come into resonance with them. If pilgrims to a holy place have been inspired, uplifted and healed there, we are more likely to have similar experiences of spiritual connection. Holy places can grow in holiness through people's experiences within them.

Two practices of pilgrimage

GO ON A PILGRIMAGE

There is no need for your pilgrimage to be expensive, or elaborate, or very time-consuming. In fact, it may be better to start with somewhere local, to get to know where you live in a new way. When you open yourself to the idea, try to feel which local holy place calls you, or at least some place that you feel is important for you.

There is a wide range of choice. In England, for instance, there are many ancient sacred places, like stone circles and long barrows; sources of rivers; holy springs and wells; venerable trees; ancient churches and great cathedrals, echoing to sacred singing and chanting almost every day. The land is literally enchanted by these perpetual choirs.

In North America there are some of the greatest sacred groves and natural sanctuaries on earth; many local wild and beautiful places; and also some great churches and cathedrals, including the cathedrals of St Patrick and St John the Divine in New York, the National Cathedral in Washington and Grace Cathedral in San Francisco, as well as powerful Roman Catholic shrines, like the Sanctuary of Chimayo, near Santa Fe in New Mexico, where many healings happen.

It is best to walk at least part of the way, if only for the last mile or two, because that makes the pilgrimage more real, more embodied.

Go with an intention, something you would like to give thanks for, or ask for, or seek inspiration for. If possible, take a pilgrim's staff with you, made from any suitable wood, such as hazel, the definitive visual emblem of the pilgrim throughout the centuries. If possible learn some songs before you set off, or pick them up from other pilgrims along the way. Sing them when you reach holy wells, ancient trees, and the goal of your journey.

When you arrive at the holy place, do not go straight in, but if possible walk around it. This circumambulation, usually clockwise, helps to make the holy place the centre. Then give an offering, maybe of flowers, as in Hindu temples, or a song, or a thanksgiving, or simply a cash donation. Then, in the holy space, you can make your prayer, and in many cathedrals and churches you can light a candle. Finally, pray for blessings on your life, on your journey home, and on those to whom you are returning.

TURN YOUR JOURNEYS INTO PILGRIMAGES

Whenever I go to a new place, I try and find the sacred centre, and go there to pay my respects. In India I go to the local temple, in Buddhist countries to a stupa or monastery, in Muslim countries to the mosque or to the shrine of a saint. In Europe and the Americas, I go to the church or cathedral that is at the centre of the community. Many Roman Catholic and Anglican churches are kept open every day, and so it is possible to go in and light a candle and say a prayer, and connect with that holy place. I find this grounds me and links me to the villages, towns and cities I am visiting, and to the people who live there, as well as providing a quiet place to become centred after a journey. I ask for blessings on my time in that city and on those who I am going to meet, as well as on my friends and family at home.

I suggest you try something like this on your journeys.

8

Conclusions:
Spiritual Practices in a Secular Age

In traditional hunter-gatherer societies, there was no distinction between religion and the rest of social and cultural life. The existence of spirits, the invisible influences of ancestors and participation in collective rituals were taken as read.

Likewise in traditional agricultural societies and ancient civilisations, everyone was included in the community's religious life, although there was often a specialised priesthood. In Europe as recently as AD 1500, practically everyone believed in God and took part in religious ceremonies, festivals, and rituals. To become an atheist or to deny the importance of religion was almost inconceivable. The same is true today in many parts of the world.

By contrast, in twenty-first-century Europe, the public space is secular. An atheist or agnostic worldview is the default position in academic, intellectual, commercial and media circles. The practice of religion is a minority pursuit, and there is also a wide plurality of religions and spiritual practices, rather than one agreed set of practices that includes almost everyone. We live in a secular age unprecedented in human history.

The word *secular* itself shares a linguistic root with the word *seed*, and its primary meaning is to do with generation. In the Middle Ages, it referred to worldly affairs – activities within the realm of time, as opposed to eternity. Within the medieval church, there was a division of labour between the religious orders of monks and nuns, who had the time and opportunity to turn their hearts and minds to God's eternity, and the priests who ministered to lay people and their worldly concerns, who were called secular priests. The same terminology is still used in the Roman Catholic Church today: monks and nuns are called religious, and parish priests secular.

But 'secular' now has far wider meanings. The long process of secularisation in Europe has roots that stretch back to the Protestant Reformation in the sixteenth century. The Reformation undermined the authority of religious institutions, practices and doctrines that almost everyone had taken for granted.

As the philosopher Charles Taylor shows in his book *A Secular Age*, in the Reformation, spiritual and magical powers were removed from the outer world, while significance and meaning were transferred to individual human minds. In the pre-Reformation world, spiritual power resided in physical objects, such as the relics of saints or the consecrated Host, as well as in people. Humans were porous. They were vulnerable and healable, open to blessings or curses, possession or grace. As Taylor put it, 'in the enchanted world, the line between personal agency and impersonal force was not at all clearly drawn.'[1] By contrast, in the post-Reformation world, objects could affect minds, but their meanings were generated by minds, or imposed on things by minds. Meaning and significance were internal, inside human minds, not in the external world. The world was disenchanted.

The rising influence of mechanistic science accelerated this process from the seventeenth century onwards. God was removed from the workings of nature, now seen as inanimate, unconscious and mechanical, functioning automatically. Some Protestant theologians responded by emphasising God's role as the creator of the world machine. As we have seen, God was like an engineer who had ordered the universe benevolently for human benefit. God also retained a role at the end of time as the Judge who distributed rewards and punishments. In this process, God was reduced to a creator and religion to morality.[2] This stripped-down form of Christianity left little place for the saving action of Christ, the role of devotion and prayer, or a transcendent goal for humanity. The traditional Christian doctrine of human participation in God's nature was eclipsed.[3]

The evangelical religious movements of the eighteenth and nineteenth centuries, most notably the Methodists, reacted against this intellectual conception of God. Instead, they offered an internalised, heart-centred faith, an intensely personal form of religion, as opposed to the formal collective observances of the Roman Catholic and Anglican Churches, which sought to include everyone.

In Roman Catholic countries, the official assumption was that people should be Roman Catholics; in Lutheran countries, Lutherans; and in England, Anglicans. By contrast, new denominations, like the Methodists, were more like affinity groups or voluntary associations.[4] They did not claim an exclusive monopoly of religious correctness, and people felt free to move from one denomination to another. The United States was born in this context, and its many denominations provided, and still provide, a kind of religious free market. In the twentieth century, Pentecostalists and other evangelical churches spread this personal form of relationship with God throughout Latin America, Africa and Asia.

By the end of the eighteenth century, for many Enlightenment intellectuals, this rational creator had become the remote God of Deism, who could be known through reason, science, and the study of nature. There was no need for Revelation, or for the practices of the Christian religion, or for the 'enthusiasm' of the evangelicals. The word *enthusiasm* means 'possessed by God', from the Greek *en* = in and *theos* = god; and for Enlightenment intellectuals it was a term of disparagement. Once the universe had been made and set in motion, it functioned automatically with no need for divine interference. God did not respond to prayers, nor did he intervene by reaching into the universe, temporarily suspending the laws of nature to bring about miracles.

But what about morality and the social order? If moral behaviour no longer depended on God's commandments, guidance or grace, then it must depend on humans themselves, on reason and on the rational ordering of mutual benefit. Christianity was based on a moral universalism, with Christ's call to care for others, and to show love to neighbours, and even to enemies. This Christian ideal was secularised into a humanist morality, whereby we *ought* to be altruistic and we *ought* to be concerned with others.[5]

These secularising changes were expressed most dramatically in the French Revolution. When the Revolution began in 1789, Catholicism was the official religion of the French state. In 1793, the Cult of Reason was proclaimed the state religion and, as we have seen, the Cathedral of Notre Dame in Paris was converted into a Temple of Reason.

One of the leading revolutionary slogans was 'Liberty, Equality, Fraternity, or Death'. At least forty thousand people were executed in the Reign of Terror (1793–4), including many priests, and the guillotine became a symbol of the revolutionary cause. Churches, monasteries and religious orders were closed down, and religious worship was forcibly suppressed.

The Reign of Terror left a sour taste in the mouth, and so the revolutionary slogan was shortened to 'Liberty, Equality, Fraternity'. This is still the official motto of the Republics of France and Haiti.

Deism soon gave way to full-blown atheism. By assuming that the universe was eternal, there was no need for the creator God of Deism. Atheism became intellectually credible, and atheist revolutionary movements, including communism, spread throughout Europe in the nineteenth century. Because the old regimes had been backed by the power of the church, the revolutionary cause was strengthened by undermining the power of religion.

Especially in Russia, where the authority of the Tsar and of the Orthodox Church had rested on God, radicals saw atheism as a necessary stance. By the 1850s, revolutionary thinkers in Russia were aiming to replace the corrupt authority of Church and Tsar with a new social and political system, but also with a new concept of mankind.[6] Through rejecting the delusions of religion – the 'opium of the people' in Marx's famous phrase – humans would be liberated into the light of science and reason.

The atheist ideology found a powerful ally in materialist science, which by the end of the nineteenth century portrayed a purpose-less, unconscious, mechanical universe, in which humans, like all life, had evolved without purpose or guidance. In this godless world, humanity would take charge of its own evolution, bringing economic development, brotherhood, health and prosperity to all mankind through Progress.

Modern secularism

There are three main ways in which secularism is expressed in the modern world. The first is political and cultural. Public spaces have been emptied of God. As Taylor puts it:

> As we function within various sphere of activity – economic, political, cultural, educational, professional, recreational – the norms and principle we follow, the deliberations we engage in, generally don't refer us to God or to any religious beliefs . . . This is in striking contrast to earlier periods, when Christian faith laid down authoritative prescriptions, often through the mouths of the clergy, which could not be easily ignored in any of these domains, such as the ban on usury.[7]

This form of secularism is not necessarily anti-religious. In the United States, the separation of Church and State, laid down in the First Amendment to the Constitution in 1791, was intended to allow religious freedom, a burning issue for early Americans, many of whom had fled from state-sponsored religious persecution in Europe. Likewise, political reform movements in Europe in the nineteenth century were more often driven by a need for toleration between different Christian churches, Roman Catholic and Protestant, than by anti-religious fervour. The increasing secular-isation of Europe also made it much easier for Jews to participate in public life and to become part of the secular world.

But some secular states were explicitly anti-religious, following the precedent of revolutionary France. In the Soviet Union, atheism became the official ideology, and children were given an anti-religious education. The state-sponsored League of the Militant Godless, which in the 1930s had more than five million members, orchestrated campaigns for the closure of churches and monas-teries, silencing church bells, suppressing religious festivals and stamping out Russian Orthodox religious practices.[8] A similar atheist ideology was imposed on the Eastern European satellites of the Soviet Union after the Second World War. In Communist China, Mao Zedong instituted a policy of state atheism in 1949.[9]

The second sense in which Europe has become increasingly secular is through the decline of religious practice and affiliation. A large minority, or even a majority, of the population says they have no religion. Although the roots of this change go back to radical intellectuals in the eighteenth century and to anti-religious political movements in the nineteenth century, the process of alien-ation from traditional religion accelerated in the second half of the twentieth century, and has continued in the twenty-first century.

The third sense of secularity is the transformation of a society from one in which practically everyone believed in God to a society in which belief in God is one option among others, and frequently not the easiest option to embrace.[10]

In most of Europe, and increasingly among young people in North America, the current default option is to be non-religious, or even anti-religious.

The ambiguities of atheism

In part, this cultural drift towards atheism is the result of the continuing efforts of evangelical atheists to convert people to their point of view. Historically, modern atheism grew out of Christianity, and as the philosopher John Gray argues, it is best seen as a Christian heresy:

> Unbelief is a move in a game whose rules are set by believers. To deny the existence of God is to accept the categories of monotheism . . . Atheism is a late bloom of the Christian passion for truth . . . Christianity struck at the root of pagan tolerance of illusion. In claiming that there is only one true faith, it gave truth a supreme value that it had not had before. It also made disbelief in the divine possible for the first time. The long-delayed consequence of Christian faith was an idolatry of truth that found its most complete expression in atheism.[11]

Gray is himself an atheist, but not a proselytising atheist, nor a Secular Humanist. But many modern atheists, like Richard Dawkins, are still on a crusade against God. They are the missionaries of an anti-religious ideology. They see themselves as heirs of the Enlightenment.

For people who are brought up within a religious family, becoming an atheist involves an enormous shift of perspective, a revolutionary change in worldview. Many contemporary atheists have made that switch themselves, by rebelling against or drifting away from their Christian or Jewish religious upbringing. Others were raised by non-religious parents, and some are third-generation secular, with non-religious grandparents. Few non-religious ancestries go back much further. I was a first–generation atheist, and went through this paradigm shift when I was a teenager.

Many atheists regard this shift from religion to atheistic secularism as historically inevitable, and to some extent this is a self-fulfilling prophecy. Those who convert to atheism, or simply to a non-religious lifestyle, often see themselves as progressive and, like Dawkins, inheritors of the Enlightenment ideal of progress. And in some ways the Enlightenment programme seems to be coming true, at least in Europe, North America, Australia and New Zealand. Public life, the educational system and the media have become increasingly secular, and the non-religious grow in numbers as churches decline.

One of the commonest atheist arguments against religion is that religions cause conflicts. This is true. The Thirty Years War (1618–48) between Catholic and Protestant states claimed over three million lives in Europe. The notorious Spanish Inquisition (1478–1835), over its 357-year history charged about 150,000 people with offences, and executed about 3,000.[12] Religions have led to violence, and some still do.

But so has nationalism led to violence, as in Nazi Germany; so has imperialism, as in the British, French, Spanish, Portuguese, Dutch and Belgian empires; and so has colonialism. The settling of the Americas, Australia, New Zealand and other parts of the world by Europeans was disastrous for the indigenous inhabitants, many of whom were killed, enslaved, dispossessed, or wiped out by diseases.

The most destructive system of all was the atheist ideology of communism, as in the Soviet Union under Stalin, Communist China under Mao and Cambodia under Pol Pot. By conservative estimates, the death toll in the Soviet Union under Stalin was about twenty million people,[13] with a further twenty million soldiers and civilians killed in the Second World War.[14] In China under Mao there were forty to seventy million deaths as a result of his policies.[15] In Cambodia under Pol Pot about two million people perished, around a quarter of the population.[16]

No nation, no religion, no ideology, and no commercial system comes out well from a close examination of its history. All human institutions are fallible.

Historical arguments about the bad deeds of religion are an important part of the atheist worldview, while ignoring or brushing aside the vast death toll of atheist regimes. But even more important is the atheist belief that science has already explained the nature of reality in purely physical terms, with no need for God. The universe itself, and living organisms are machines. They have evolved automatically and unconsciously, with no creator, no creative intelligence, and no purpose.

But this 'scientific worldview', the materialist theory of nature, rests on assumptions that are all highly questionable scientifically, as I show in my book *The Science Delusion* (called *Science Set Free* in the US). For example, materialists have not *proved* that matter is unconscious, or that nature is purposeless, or that minds are confined to brains. These are *assumptions*. The materialist worldview is a belief system, not a statement of scientific facts.

Another common reason for converting to atheism is the assumption that religions are primarily about propositions and beliefs, rather than about experiences. Then religions can be dismissed as dogmatic, dependent on the authority of scriptures, prophets and priests. By contrast, so the argument goes, scientists are open to evidence; they ask clear questions, test them by experiment and establish a reliable consensus through repeatable observations.

I used to believe this myself. But I was disillusioned when I found that some people have made science into a kind of religion, and are often exceptionally dogmatic. They accept the 'scientific worldview' on faith, impressed by the authority and prestige of scientists, and imagine that they have arrived at this worldview by their own free thinking. I still believe in the ideal of open-minded science. But I see the religion of science, scientism, as a dogmatic ideology. In my own experience, believers in scientism are much more dogmatic than most Christians I encounter.

Most believers in scientism are not scientists. They are devotees rather than researchers. Most have made no empirical observations or scientific discoveries themselves. They have not worked at the Large Hadron Collider studying subatomic particles, nor have they sequenced genomes, nor examined the ultrastructure of nerve cells, nor done research in radio astronomy, nor penetrated the mathematics of superstring theory. They take what they are told on trust, accepting the prevailing orthodoxy of institutional science, as conveyed by textbooks and popularisers. They are incapable of questioning the authority of the scientific priesthood because they lack the necessary education and technical knowledge to do so. And if they do raise awkward questions, they are likely to be ignored, or dismissed as ignorant, confused or stupid.

Scientism has a very wide influence, largely because of the undoubted triumphs of science and technology, such as computers, the Internet, smartphones, antibiotics, keyhole surgery, jet engines and space probes. It is easy to assume that all these triumphs are a result of the 'scientific worldview', and that they support the materialist philosophy of nature. But many people, including me, think that this philosophy has hardened into a dogmatic belief system that is actually holding the sciences back.[17]

The least successful aspect of the modern sciences is in the understanding of consciousness. Materialists assume that it is nothing but the activity of brains. Their slogan is, 'Minds are what brains do.' But the very existence of consciousness is a problem for materialists, and is often called the 'hard problem' (as discussed in Chapter One).

Religions are about consciousness, and are founded on the assumption that consciousness transcends the human level. That is one reason why believers in scientism are anti-religious; all religions assume that consciousness is more extensive than brain activity.

When people give up their ancestral religion, they stop most of the practices that religious people take for granted, including singing and chanting together, praying, participating in traditional rituals and festivals, and saying grace before meals. What are the effects of this paradigm shift from a religious to a non-religious way of life?

The effects of religious and non-religious ways of life

Experimental scientific research on spiritual practices takes place in a secular context. The researchers usually assume that the participants are not religious and that they have no spiritual practices. They then investigate the effects of adding a particular practice and study its effects, compared with a control group that does not take part in this practice. For instance, research on gratitude compares the effect of expressing gratitude with not expressing gratitude (Chapter Two). Research on meditation compares the effect of meditating with not meditating (Chapter One). Research on spending time outdoors compares being outdoors with being indoors, the default situation (Chapter Three). Research on the effects of singing compares singing with not singing (Chapter Six). Most of these studies show that spiritual practices have beneficial effects, compared with not taking part in these practices.

Another way of investigating the effects of spiritual practices is to look at the long-term effects of religious participation, as opposed to non-participation. People who regularly go to churches or synagogues, or other places of religious assembly, are compared with people who do not do so, matched for similar age-ranges, economic and social status. Thousands of these studies have been carried out in the US and elsewhere. The findings are clear.

People who regularly went to church tended to have less mental illness, suffer less depression, show less anxiety, and live longer than those with little or no religious participation.[18] This effect was not confined to Christianity. There was a similar effect in Taiwan in a predominantly Buddhist context.[19]

There are exceptions. For a minority of people, especially those who are filled with guilt, or fear, or who have experienced severe religious conflicts, religious beliefs can have negative effects on health and wellbeing.[20]

But most people who have given up their ancestral faith have not done so to escape extreme guilt or conflict. Many convert to a non-religious lifestyle not for negative but for positive reasons, as discussed above; they want to align themselves with progress, reason and science.

When people abandon their ancestral religion, they usually cease to take part in a whole series of practices that their ancestors took for granted, including:

- Giving thanks as a community and as a family
- Being part of a community that sings together
- Praying
- Accepting death as a transition and not an end
- Going through rites of passage like baptism, confirmation and marriage
- Having traditional religious funerals
- Acknowledging ancestors
- Celebrating festivals that give structure to the passage of time through the year
- Connecting with sacred places
- Participating in rituals that give a sense of collective identity and continuity
- Being encouraged to help other people
- Being part of a larger story that helps give meaning to individual lives
- Feeling a connection to a spiritual reality that transcends space and time

By giving up religion and the practices that go with it, people's lives are indeed freed from restrictions. For ex-Christians, Sunday need no longer be a special day of thanksgiving, rest and recreation; it can be just another day for work or shopping. There is no religious barrier to a 24/7 lifestyle.

Much changes in this process, not only for first-generation non-religious people, but also for their children. Unlike children of religious families, children of non-religious families do not as a matter of course sing with their family's community, and give thanks together, and take part in rituals and festivals.

Atheism is a purifying fire. It burns up religious hypocrisy, corruption, laziness and pretention. But its scorched-earth policy can leave many people spiritually hungry, thirsty and isolated.

Over the last few generations, this paradigm shift from a religious to a non-religious way of life has happened on a vast scale in previously Christian countries in Europe, North America, Australia and New Zealand. But the abandonment of religion has not involved a full-scale conversion to atheism. Most people who have given up their family's religion, or who are raised in a non-religious family, do not see themselves as atheists.[21] Some call themselves agnostics; others still retain a tenuous religious affiliation, like going to church at Christmas; others are spiritual seekers; others are New Agers; others adopt some of the practices of other religions, like Buddhism, or become neo-pagans or neo-shamans.

In recent surveys in the UK, about half the population said they had no religion,[22] but only thirteen per cent described themselves as atheists. Even among the non-religious, only twenty-five per cent agreed with the statement that 'humans are purely material beings with no spiritual element.'[23]

The proportion of atheists is higher among the scientifically educated. According to a survey in the UK in 2016, among scientific, engineering and medical professionals, about twenty-five per cent were atheists, and twenty-one per cent agnostics, making forty-six per cent altogether. An almost equal proportion, forty-five per cent, said they belonged to a religion or that they were spiritual but not religious.[24] Thus in the UK, one of the most secular countries in the world, even in the scientific community, fully fledged atheists are in a minority.

In general, religious and spiritual practices make people happier, healthier and less depressed. Conversely, not having these practices makes people unhappier, unhealthier and more depressed. Militant atheism should come with a health warning.

Some atheists recognise this problem, which is why Alain de Botton advocates religion for atheists. This is why Secular Humanists train and license humanist officiants to carry out secular naming ceremonies, weddings and funerals. This is why the Sunday Assembly provides weekly opportunities for group singing. This is why Sam Harris and secular Buddhists advocate meditation. A rigorously non-religious lifestyle leaves too much out, impoverishing people's lives.

Spiritual practices as ways of connecting

The advantage of most spiritual practices is precisely that they are about practice rather than belief. They are therefore open to religious people and to non-religious people. They are inclusive.

Spiritual practices take us beyond our immediate concerns. At first sight, the practices discussed in this book relate to very diverse aspects of human experience. What is the common thread?

Connection is the theme that unifies them all. They lead us beyond the mundane to deeper kinds of connection:

1. **Gratitude** is about the flow of giving and receiving. Being part of a flow connects us. We can choose how far we go in acknowledging and giving thanks. In the human realm, we can feel

grateful to all those who have helped us and sustained us, including our parents, who gave us our life. In the more-than-human world we can give thanks for other living organisms that surround us and on which we depend for our survival, and for the life of the earth. We can go further and give thanks for the sun, the galaxy and the entire cosmos. We can go further still, and feel grateful to the source of all nature and all minds, whether we call this God or not.

We are free to be as grateful or as ungrateful as we like. The more ungrateful we are, the greater our disconnection, discontent and isolation. The more grateful we are, the deeper our connection with a greater life than our own, and the stronger our experience of flow. This consciousness of flow helps us to be more giving, more generous.

2. **Meditation** makes us aware of the activities of our minds, as we see them drawing us into a process in time, connecting our past to our future in personal tunnels. Through meditation we can step back to a more inclusive consciousness. And sometimes we find ourselves in a vastly greater presence of mind, a mind far beyond our own. We are connected through conscious presence.

3. **Connecting with the more-than-human world.** We can go as far as we choose with our minds and our senses. We can pay attention to the world of animals, plants, fungi, microbes, forests, oceans, the weather, Gaia, the sun, the solar system, the Milky Way, and countless galaxies beyond our own. We can reach towards the source from which all nature comes.

4. **Plants** offer us connections to life forms totally different from our own. Like us, plants grow and become. But unlike plants, we stop growing and start behaving, as do other animals.

Plants are the source of qualities that we and other animals experience: forms, smells, tastes, textures and colours. They feed us, directly or directly; they heal us as herbs, or poison us. Some drug-containing plants can change our minds. And they are much older than we are. The main families of flowering plants have been around for tens of millions years, conifers for 300 million years, ferns, mosses, seaweeds and other algae even longer. Our species is only about 0.1 million years old, and civilisation only about 0.005 million years old.

5. **Rituals** connect us with those who have performed the rituals before. They maintain the traditions and continuity of our group, as well as opening a channel to more-than-human consciousness. Rituals also connect us with our descendants and all who will perform the rituals again. Through rituals we are connected to the past and the future of our group, and to the spiritual realm to which our group is linked and to a transcendent goal for humanity.

6. **Singing, chanting and music** link members of the group in synchrony and resonance. Mantras, chants, songs and dances can connect us to the more-than-human world and to more-than-human minds. Music links us to the flow of life.

7. **Pilgrimage** connects us to holy places, places where heaven and earth are joined. In many holy places this is literally true. Their structures reach up into the sky as standing stones, obelisks, towers, spires and minarets.

Pilgrimage has the great advantage of being both a practice and a metaphor. Through going on a pilgrimage, we experience the process of moving towards the goal and arriving at it, being there. Then we go home changed. We connect our ordinary everyday lives with places that link us to a transcendent world.

We can see our entire life as a pilgrimage. Depending on our beliefs, this can either be a journey whose destination is our inevitable death, or a journey towards a spiritual connection at the hour of our death, as in a Near Death Experience, and a journey that continues beyond our death.

Journeys of discovery and rediscovery

There are many spiritual practices, and all religions include a wide range of them. These are not mutually exclusive; they are mutually reinforcing.

The seven spiritual practices I have discussed in this book are by no means an exhaustive catalogue, and in a sequel to this book I hope to discuss a range of other practices, including prayer, fasting, psychedelics, and holy days and festivals.

Not all practices work equally well for everyone, and all of us have to make our own choices among them. For people who follow a religion, many of these practices are already part of their lives. But often their effectiveness is dulled through familiarity. By looking at these practices afresh, their power can be renewed.

Each religious path involves its own selection of spiritual practices, emphasising some more than others. As a result, some are unfamiliar to people who are already following a religious path. For example, many Protestant Christians are not used to going on pilgrimages. These sacred journeys were familiar to their pre-Reformation ancestors, and are still familiar in the Eastern Orthodox and Catholic Churches. Likewise, contemplative prayer, and other Christian forms of meditation are well known in communities of monks and nuns, but less well known among lay people, who often benefit from discovering them.

One of the areas in which religious people can learn from the non-religious is in connecting with the more-than-human world in new ways opened up by science. Even the most atheistic scientists form a relationship with the natural world through their investigation of it, however specialised their field of study. Many religious people lack this sense of connection with the details of nature, and some seem impatient to soar beyond them.

This is an area with a huge potential for spiritual exploration. The natural sciences have unveiled a universe far larger, older and stranger than anything previously imagined; they have revealed details about biological life that no one knew before; they have unveiled the existence of realms of micro-organisms around us, and also within us: the vast community of microbes that lives in our guts. The sciences have penetrated into realms of the very large and the very small which our ancestors knew nothing about. The trouble is that the sciences give us vast amounts of data, but it is devoid of personal or spiritual meaning.

By contrast, traditional spiritual connections with the more-than-human world found meaning and significance everywhere; but knew nothing of these recent discoveries of the sciences. To combine these two approaches is a uniquely modern challenge.

We are all on journeys. Spiritual practices can enrich our lives and give us a stronger sense of connection with each other, as well as with life and consciousness beyond the human level. These practices can help us accept some of the many gifts that we are offered, and give thanks for them. The more we appreciate what we have been given, the greater our motivation to give.

Acknowledgements

In thinking about the themes I discuss in this book, I have been greatly helped by my discussions with Father Bede Griffiths, with my wife Jill Purce, and with many other people. In particular, I had the good fortune to take part in a series of trialogues (three-way dialogues) with my friends Terence McKenna (who, sadly, died in 2000) and Ralph Abraham, over a period of seventeen years. We met together at least once a year in California, England or Hawaii, and discussed a wide variety of subjects, some closely related to subjects discussed in this book. We also published two books together, *The Evolutionary Mind* (1998) and *Chaos, Creativity and Cosmic Consciousness* (2001).[1] Recordings of more than thirty of our trialogues are available online.[2]

I have learned a great deal from dialogues with several spiritual leaders and teachers, including Jiddu Krishnamurti;[3] Brother David Steindl-Rast, with whom I led workshops at the Esalen Institute in California and at Hollyhock, on Cortes Island in British Columbia; Matthew Fox, with whom I wrote two books on science and spirituality, *Natural Grace* (1996) and *The Physics of Angels* (1996), led workshops at Hollyhock and in Oakland, California,[4] and made a series of podcasts;[5] Marc Andrus, the Bishop of California, with whom I led workshops at Esalen and at Grace Cathedral, San Francisco, and with whom I have also made a series of podcasts;[6] Mark Vernon, with whom I have made more than thirty podcasts;[7] David Abram, Rick Ingrasci and Stephen Tucker, the vicar of my parish church in Hampstead. I am grateful to them all. I also thank my sons Merlin and Cosmo, with whom I have led workshops at Hollyhock over the last four years exploring some of the themes discussed in this book. The feedback and responses from participants in these various workshops have been invaluable.

I am thankful for the financial support that has enabled me to write this book: from Addison Fischer and the Planet Heritage Foundation, of Naples, Florida and the Gaia Foundation in London; from Ian and Victoria Watson, the Watson Family Foundation and the Institute of Noetic Sciences, in Petaluma, California.

I thank Pam Smart, my Research Assistant, who has worked with me for twenty-two years; Guy Hayward, my Post-Doctoral Research Fellow, who has helped with research for this book; and Sebastian Penraeth, my webmaster.

I much appreciate the encouragement of my editor, Mark Booth, at Hodder & Stoughton, in London, who has helped make this book a reality. I also thank all those who have commented on drafts of this book, especially Angelika Cawdor, Lindy Dufferin and Ava, Guy Hayward, Natuschka Lee, Will Parsons, Jill Purce, Anthony Ramsay, Cosmo and Merlin Sheldrake and Pam Smart.

Notes

Preface

1. Heller, 1952.
2. http://epiphanyphilosophers.org. Retrieved 17 February 2017.
3. Braithwaite, 1953.
4. See for example Griffiths, 1976, 1982.
5. http://www.jillpurce.com. Retrieved 1 February 2017.
6. Sheldrake, 2009, 2011.
7. Sheldrake, 2002.
8. Sheldrake, 1999.
9. Sheldrake, 2003.
10. Sheldrake and Smart, 2003.

Introduction

1. Koenig et al. 2012, chaps. 7, 9.
2. Ibid., chap. 7.
3. Ibid., chap. 26.
4. Ibid., chap. 11.
5. http://www.pewforum.org/2011/12/19/global-christianity-exec/. Retrieved 11 November 2016.
6. http://www.pewforum.org/2014/02/10/russians-return-to-religion-but-not-to-church/. Retrieved 25 November 2016.
7. For a magisterial and insightful discussion, see: Taylor, 2007.
8. http://www.brin.ac.uk/figures/. Retrieved 8 November 2016.
9. http://www.lancaster.ac.uk/news/articles/2016/why-no-religion-is-the-new-religion/. Retrieved 8 November 2016.
10. http://about-france.com/religion.htm. Retrieved 8 November 2016.

11. http://www.pathwaystogod.org/resources/thinking-faith/religious-landscape-sweden. Retrieved 8 November 2016.

12. http://www.irishcentral.com/news/numbers-in-irelands-catholic-church-continue-to-drop-stigma-attached-to-attending-mass-200315991-237575781. Retrieved 16 November 2016.

13. http://worldnews.nbcnews.com/_news/2013/03/05/17184588-as-church-attendance-drops-europes-most-catholic-country-seeks-modern-pope?lite. Retrieved 8 November 2016.

14. Hout, M. and Smith, T.W., 'Fewer Americans affiliate with organised religions, belief and practice unchanged', Key findings from the 2014 General Social Survey, 2015: http://www.norc.org/PDFs/GSS%20Reports/GSS_Religion_2014.pdf /. Retrieved 8 November 2016.

15. http://www.pewforum.org/2015/11/03/u-s-public-becoming-less-religious/. Retrieved 8 November 2016.

16. Theos, 2013.

17. For some striking examples, see Douthat, R., 'Varieties of religious experience', *New York Times*, 24 December 2016: http://www.nytimes.com/2016/12/24/opinion/sunday/varieties-of-religious-experience.html?_r=0. Retrieved 27 December 2016.

18. http://www.philosophyforlife.org/the-spiritual-experiences-survey/. Retrieved November 2016.

19. Ibid.

20. Woodhead and Catto, 2012.

21. http://www.philosophyforlife.org/the-spiritual-experiences-survey/: Retrieved 9 November 2016.

22. De Botton, 2013, p. 13.

23. Ibid., p. 14.

24. Harris, 2014, pp. 202–3.

25. https://www.samharris.org/blog/item/how-to-meditate. Retrieved 9 November 2016.

26. http://www.sundayassembly.com. Retrieved 9 November 2016.

27. http://www.philosophyforlife.org/category/atheism/. Retrieved 9 November 2016.

28. Koenig et al., 2001.

29. Koenig et al., 2012.

Chapter 1: Meditation and the Nature of Minds

1. Benson and Klipper, 2000.

2. Simons, N., 'MPs slow the Westminster treadmill with weekly "mindfulness" meetings', *Huffington Post*, 4 November 2013: http://www.huffingtonpost.co.uk/2013/10/30/chris-ruane-parliament-mindfulness_n_4177609.html. Retrieved 26 October 2016.

3. NHS Choices, Mindfulness, 2016: http://www.nhs.uk/Conditions/stress-anxiety-depression/Pages/mindfulness.aspx. Retrieved 26 October 2016.

4. Partridge, 1966, p. 393.

5. Miller, K., 'Archaeologists find earliest evidence of humans cooking with fire', *Discover*, May 2013: http://discovermagazine.com/2013/may/09-archaeologists-find-earliest-evidence-of-humans- cooking-with-fire. Retrieved 19 September 2016.

6. Taylor, 1997.

7. Ibid.

8. Chopra, D., 'The Maharishi Years – the untold story: recollections of a former disciple', *Huffington Post*, 17 November 2011: http://www.huffingtonpost.com/deepak-chopra/the-maharishi-years-the-u_b_86412.html. Retrieved 17 September 2016.

9. See for example the World Community for Christian Meditation: http://wccm.org. Retrieved 16 September 2016.

10. Benson and Klipper, 2000.

11. Kuyken et al., 2015.

12. Blackmore, 2011.

13. Harris, 2014.

14. NIH, 'Nationwide survey reveals widespread use of mind and body practices', https://www.nih.gov/news-events/news-releases/nationwide-survey-reveals-widespread-use-mind-body-practices, 2015. Retrieved 5 October 2016.

15. Benson and Klipper, 2004.

16. Ibid., pp. 65–82.
17. Ibid., pp. xxi–xxii.
18. Quoted in Fox, 2014, p. 55.
19. Koenig et al., 2001, 2012.
20. Benson and Klipper, 2004, p. xlii.
21. Schwartz, 2011.
22. Rosenthal et al., 2011.
23. Wood, D., 'Veterans find comfort in meditation therapy', *Huffington Post*, 3 March 2015: http://www.huffingtonpost.com/2015/02/20/vets-ptsd-meditation_n_6714544.html. Retrieved 4 October 2016.
24. Goyal et al., 2014.
25. Britton, W., 'The dark knight of the soul', *The Atlantic*, 15 June 2014: http://www.theatlantic.com/health/archive/2014/06/the-dark-knight-of-the-souls/372766/. Retrieved 5 October 2016.
26. Booth, R., 'Mindfulness therapy comes at a high price for some, say experts', *Guardian*, 25 August 2014: Retrieved 5 October 2014.
27. Stahl, J.E. et al., 'Relaxation Response and Resiliency Training and Its Effect on Healthcare Resource Utilization', *PLOS One*, 13 October 2015: http://journals.plos.org/plosone/article?id=10.1371/journal.pone.0140212. Retrieved 4 October 2016.
28. Lutz et al., 2004.
29. Hölzel et al., 2008.
30. Schulte, B., 'Harvard neuroscientists: meditation not only reduces stress, here's how it changes your brain', *Washington Post*, 26 May 2015: https://www.washingtonpost.com/news/inspired-life/wp/2015/05/26/harvard-neuroscientist-meditation-not-only-reduces-stress-it-literally-changes-your-brain/. Retrieved 22 October 2016.
31. Hölzel et al., 2011.
32. Mascaro, 1965, p. 51.
33. Aquinas, 2009, p. 226.
34. Batchelor, 2017.
35. Harris, 2014, p. 10.
36. Ibid., p. 137.

37. Ibid., pp. 135, 137.

38. Powers, B., 'The nondual realization of Sam Harris: the future of an illusion', 2014: http://www.integralworld.net/powers17.html. Retrieved 25 October 2016.

39. Harris, 2014, pp. 175–6.

40. https://www.samharris.org/blog/item/how-to-meditate, Retrieved 26 October 2016.

41. Wax, 2016.

42. e.g. http://www.awakenedheartproject.org. Retrieved 26 October 2016.

43. e.g. http://wccm.org. Retrieved 26 October 2016.

44. e.g. http://islamicsunrays.com/islamic-meditation-for-relaxation-and-spiritual-comfort/. Retrieved 26 October 2016.

Chapter 2: The Flow of Gratitude

1. Emmons and Crumpler, 2000.

2. McCullough et al., 2002.

3. Emmons and Kneezel, 2005.

4. Watkins et al., 2009.

5. Emmons and McCullough, 2003.

6. Bobo et al., 2004.

7. Ibid.

8. Seligman, 2005.

9. Ehrenreich, 2009.

10. Ibid.

11. Mauss, 2000, p. 20.

12. Partridge, 1966.

13. Sacks, 2015.

14. Interestingly, Oliver Sacks's nephew, Jonathan Sacks, was Chief Rabbi of Great Britain.

15. A very helpful book on this subject is *Gratefulness, The Heart of Prayer* by my friend Brother David Steindl-Rast, a Benedictine monk. He also founded an inspiring website called www.gratefulness.org.

16. Hart, 2013.
17. There are several websites where you can look up graces, or learn sung graces. One is: www.graces.io.

Chapter 3: Reconnecting with the More-Than-Human World

1. Abram, 1997.
2. Descola, 2013, p. 392.
3. Viveiros de Castro, 2004.
4. Wilson, 1984.
5. Howell, 2016.
6. For example, through the UK Conservation Volunteers: http://www.tcv.org.uk. Retrieved 12 February 2017.
7. Reynolds, 2015.
8. Gilbert, 2016.
9. Park et al., 2010.
10. Li, 2010.
11. Karjalainen et al., 2010.
12. Bratman et al., 2015.
13. Bratman, Hamilton et al., 2015.
14. Gilbert, 2016.
15. H.M. Government White Paper, 2011.
16. Louv, 2008.
17. https://www.nlm.nih.gov/medlineplus/ency/patientinstructions/000355.htm. Retrieved 29 March 2016.
18. http://www.bbc.co.uk/news/education-19870199. Retrieved 29 March 2016.
19. Hardy, 1979.
20. Ibid., p. 108.
21. Ibid., p. 49.
22. Ibid., p. 33.
23. Paffard, 1973, p. 117.
24. Ibid., p. 184.
25. Ibid., pp. 121–2.
26. Sheldrake, 1992.

27. Frazer, 1918.

28. Bentley, 1985.

29. Berresford, 1985.

30. https://westernmystics.wordpress.com/2015/03/22/hildegard-of-bingen/. Retrieved 16 March 2017.

31. Hart, 2013.

32. Eire, 1986, p. 224.

33. Roszak, 1973; Berman, 1984.

34. Thomas, 1984.

35. Ibid., p. 257.

36. Ibid., p. 258.

37. Ibid., p. 266.

38. Ibid., p. 267; Southey, 1807.

39. Emerson, 1985, pp. 38–9.

40. Wroe, 2007.

41. Darwin, 1794–6.

42. Ibid., p. 36.

43. Bowler, 1984, p. 134.

44. Darwin, 1875, pp. 7–8.

45. Darwin, 1859, chap. 3.

46. See the discussion in Sheldrake, 2012, chap. 4.

47. Strawson, 2006.

48. Nagel, 2012.

49. Strawson, 2006.

50. Ibid., p. 27.

51. This meeting was generously funded by the Lifebridge Foundation of New York, and I am grateful to them for making this gathering possible.

52. http://sohowww.nascom.nasa.gov/spaceweather/. Retrieved 16 September 2016.

53. Anselm, 2008, chap. 16.

54. Young et al., 2010.

Chapter 4: Relating to Plants

1. Cambridge University Frank Smart Prize for Botany, 1962.
2. http://www.sheldrake.org/research/plant-and-cell-biology. Retrieved 30 January 2017.
3. Darwin, 1882, p. 185.
4. Ibid., p. 186.
5. Catholic Church, 1999, p. 17.
6. Hart, 2013.
7. Anselm, 2008, p. 97.
8. Deb, 2007.
9. Frazer, 1918, vol. 3, p. 62.
10. Anderson and Hicks, 1990.
11. Quoted at https://www.walden.org/Library/About_Thoreau's_Life_and_Writings:_The_Research_Collections/Thoreau_and_the_Environment. Retrieved 28 January 2016.
12. Williams, 2002.
13. Giblett, 2011.
14. Ibid., p. 143.
15. Ibid., p. 145.
16. https://en.wikipedia.org/wiki/List_of_Sites_of_Special_Scientific_Interest_by_Area_of_Search. Retrieved 11 November 2016.
17. http://www.garden.org/about/press/press.php?q=show&pr=pr_nga&id=3819. Retrieved 30 March 2016.
18. http://www.dickiesstore.co.uk/blog/2014/04/16/infographic-gardening-uk. Retrieved 16 March 2016.

Chapter 5: Rituals and the Presence of the Past

1. Strehlow, quoted in Lévi-Strauss, 1972, p. 235.
2. Op. cit., p. 236.
3. Eliade, 1958, p. 391.
4. Sheldrake and Fox, 1996.
5. van Gennep, 1972.
6. La Fontaine, 1985.
7. See for example, http://schooloflostborders.org/content/huffington-

post-what-vision-quest-and-why-do-one. Retrieved 17 February 2017.

8. Corazza, 2008.
9. http://iands.org/ndes/about-ndes.html. Retrieved 15 June 2016.
10. Carter, 2010.
11. http://iands.org/ndes/about-ndes/common-aftereffects.html. Retrieved 15 June 2016.
12. van Lommel, 2011.
13. Schweiker, W., 'Torture and religious practice', 2008. http://onlinelibrary.wiley.com/doi/10.1111/j.1540-6385.2008.00395.x/full. Retrieved 15 June 2016.
14. Ibid.
15. Freud, 1939, p. 95.
16. McDougall, D., 'Indian cult kills children for goddess.' https://www.theguardian.com/world/2006/mar/05/india.theobserver. Retrieved 17 June 2016.
17. Paye-Layleh, J. 'I ate children's hearts, ex-rebel says.' http://news.bbc.co.uk/1/hi/world/africa/7200101.stm. Retrieved 20 March 2017.
18. Ehrenreich, B., 1997.
19. Ibid., p. 40.
20. Ibid., p. 59.
21. Ibid., p. 41.
22. Ibid., p. 54.
23. http://www.humanesociety.org/issues/biomedical_research/qa/questions_answers.html. Retrieved 17 June 2016.
24. http://www.slate.com/blogs/lexicon_valley/2015/07/09/the_surprising_history_of_scientific_researchers_using_the_word_sacrifice.html. Retrieved 17 June 2016.
25. http://www.lablit.com/article/394. Retrieved 17 June 2016.
26. Sheldrake, 2012.
27. Sheldrake, 1988.
28. Sheldrake, 2009, 2011.
29. Woodard and McCrone, 1975.
30. Quoted in Woodard and McCrone, 1975.
31. Details are given in Sheldrake, 2009, 2011.

32. Ibid.
33. Sheldrake, 2009.
34. See the discussion in Sheldrake and Fox, 1996, chap. 6.
35. https://www.psychologytoday.com/blog/hide-and-seek/201402/the-history-kissing. Retrieved 7 July 2016.

Chapter 6: Singing, Chanting and the Power of Music

1. www.healingvoice.com. Retrieved 31 December 2016.
2. Sacks, 2007, pp. x–xi.
3. Darwin, 1885, p. 567.
4. Ibid., p. 569.
5. Cross, 2016.
6. Ibid., p. 11.
7. Darwin, 1885, p. 572.
8. Ibid., p. 571.
9. Brown, 2004.
10. Cross and Morley, 2008.
11. Merker, 2000.
12. Geissmann, 2000.
13. Zivotofsky et al., 2012.
14. Brown, 2004.
15. Cross and Morley, 2008.
16. Ibid.
17. Bellah, 2011.
18. www.healingvoice.com. Retrieved 13 January 2017.
19. Purce, 1986.
20. Ibid.
21. For a detailed research study see Hayward (2014): https://cambridge.academia.edu/GuyHayward. Retrieved 20 February 2017.
22. Sheldrake, 2011.
23. Clift et al., 2008.
24. Clift et al., 2010.
25. Chanda and Levitin, 2013.
26. Clift et al., 2010.

27. Vella-Burrows, 2012.
28. Clift et al., 2010.
29. Chanda and Levitin, 2013.
30. Ibid.
31. Levitin, 2006.
32. Chanda and Levitin, 2013, p. 186.
33. Levitin, 2006, p. 173.
34. Ibid., p. 191.
35. Feldman et al., 2016.
36. Chanda and Levitin, 2013.
37. Ibid.
38. Nilsson, 2009.
39. It was her awareness of a correlation between the degree of apparent development in societies and their lack of group singing and chanting that originally inspired Jill Purce to bring people together to rediscover the power of group chant.
40. Frazer, 1918.
41. Ehrenreich, 2006, p. 53.
42. Ibid., p. 65.
43. Ibid., p. 137.
44. Ibid., p. 4.
45. http://www.everydayhealth.com/hs/major-depression/depression-statistics/. Retrieved 20 January 2017.
46. Quoted Irvin, J. and McLear, C. (eds), in *The Mojo Collection* (4th ed.), Canongate Books, 2003, p. 20.
47. Quoted in Godwin, 1986, p. 6.
48. Ibid., pp. 10–11.
49. Titze and Worley, 2008.
50. Hazrat Inayat Khan, 2009.
51. Tolkien, 2013.

Chapter 7: Pilgrimages and Holy Places

1. https://50societyx.wordpress.com/2013/02/12/the-dance-of-the-kingfish/. Retrieved 3 February 2017.

2. http://www.nature.com/articles/srep22219. Retrieved 3 February 2017.

3. Vitebsky, 2011.

4. Boyles, 1991.

5. Michell, 1975, p. 10.

6. Ibid.

7. Frazer, 1918, vol. 2, p. 76.

8. Coleman and Elsner, 1995.

9. Ibid.

10. Ibid., p. 20.

11. Ibid., p. 25.

12. Brown, 2015.

13. Coleman and Elsner, 1995.

14. http://www.york.ac.uk/projects/pilgrimage/content/reform.html. Retrieved 15 June 2016.

15. Nabokov, 2006.

16. Albera and Eade, 2015.

17. Gray, 2011.

18. https://en.wikipedia.org/wiki/Lenin%27s_Mausoleum. Retrieved 25 January 2016.

19. Davies, 1998, p. 80.

20. Partridge, 1966, pp. 663–4

21. http://www.statista.com/topics/962/global-tourism/. Retrieved 27 January 2016.

22. Davidson and Gitlitz, 2002.

23. http://www.csj.org.uk/the-present-day-pilgrimage/thoughts-and-essays/2000-years-of-the-pilgrimage/. Retrieved 27 January 2016.

24. Ibid.

25. https://oficinadelperegrino.com/estadisticas/. Retrieved 15 January 2016.

26. http://pilegrimsleden.no/en/. Retrieved 27 January 2016.

27. http://britishpilgrimage.co.uk/the-bpt/. Retrieved 27 January 2016.

28. e.g. Morris, 1982.

29. https://www.walkingforhealth.org.uk/sites/default/files/

Walking%20works_LONG_AW_Web.pdf. Retrieved 30 January 2016.

30. http://www.mindingourbodies.ca/about_the_project/literature_reviews/the_nurture_of_nature. Retrieved 30 January 2016.

31. http://ajot.aota.org/article.aspx?articleid=1862485. Retrieved 30 January 2016.

32. e.g. http://www.health.harvard.edu/mind-and-mood/exercise-and-depression-report-excerpt. Retrieved 30 January 2016.

33. Evans, 2010.

34. https://en.wikipedia.org/wiki/Lourdes. Retrieved 30 January 2016.

35. http://en.lourdes-france.org/deepen/cures-and-miracles. Retrieved 30 January 2016.

36. Quoted in Michell, 1975, p. 13.

37. http://www.mcofs.org.uk/lightning.asp. Retrieved 26 January 2016.

38. Scott et al., 2014.

39. http://www.lightningmaps.org. Retrieved 3 February 2017.

Chapter 8: Conclusions: Spiritual Practices in a Secular Age

1. Taylor, 2007, p. 32.

2. Ibid., p. 225.

3. For a helpful summary of Taylor's arguments, see Smith, 2014.

4. Taylor, 2007, p. 449.

5. Ibid., p. 56.

6. Spencer, 2014.

7. Taylor, 2007, p. 2.

8. http://www.encyclopedia.com/history/encyclopedias-almanacs-transcripts-and-maps/league-militant-godless. Retrieved 31 December 2016.

9. However, since 1978 the constitution of the People's Republic of China has guaranteed freedom of religion.

10. Taylor, 2007, p. 5.

11. Gray, 2002, pp. 127–8.

12. https://en.wikipedia.org/wiki/Spanish_Inquisition. Retrieved 23 December 2016.
13. Gray, 2011, p. 182.
14. http://www.ibtimes.com/how-many-people-did-joseph-stalin-kill-1111789. Retrieved 2 January 2017.
15. http://www.ibtimes.com/how-many-people-did-joseph-stalin-kill-1111789. Retrieved 2 January 2017.
16. http://www.historyplace.com/worldhistory/genocide/pol-pot.htm. Retrieved 2 January 2017)
17. Sheldrake, 2012.
18. Koenig, 2008.
19. Ibid., p. 142.
20. Ibid.
21. Zuckerman, 2007.
22. Field, 2015.
23. Theos, 2013.
24. Lorimer, 2017.

Acknowledgements

1. Sheldrake, McKenna and Abraham, 1998, 2001.
2. http://www.sheldrake.org/audios/the-sheldrake-mckenna-abraham-trialogues. Retrieved 1 February 2017.
3. http://www.sheldrake.org/videos/the-nature-of-the-mind-a-discussion-between-j-krishnamurti-david-bohm-john-hidley-and-rupert-sheldrake. Retrieved 1 February 2017.
4. Sheldrake and Fox, 1996; Fox and Sheldrake, 1996.
5. http://www.sheldrake.org/audios/discussions-between-rupert-sheldrake-and-matthew-fox. Retrieved 17 February 2017.
6. http://www.sheldrake.org/audios/dialogues-with-bishop-marc-andrus-at-grace-cathedral. Retrieved 1 February 2017.
7. http://www.sheldrake.org/audios/science-set-free-podcast. Retrieved 1 February 2017.

Bibliography

Abram, D., *The Spell of the Sensuous: Perception and Language in a More-Than-Human World*, Vintage, New York, 1997.

Albera, D. and Eade, J. (Eds.), *International Perspectives on Pilgrimage Studies: Itineraries, Gaps and Obstacles*, Routledge, New York, 2015.

Anderson, W. and Hicks, C., *Green Man: The Archetype of Our Oneness with the Earth*, HarperCollins, London, 1990.

Anselm, *Proslogion*, in: *Anselm of Canterbury: The Major Works*, Oxford University Press, 2008.

Aquinas, T., *Compendium of Theology*, Oxford University Press, 2009.

Batchelor, S., *Secular Buddhism: Imagining the Dharma in an Uncertain World*, Yale University Press, 2017.

Bellah, R.N., *Religion in Human Evolution: From the Paleolithic to the Axial Age*, Harvard University Press, 2011.

Benson, H. and Klipper, M.Z., *The Relaxation Response* (new edition), HarperTorch, New York, 2000.

Bentley, J., *Restless Bones: The Story of Relics*, Constable, London, 1985.

Berresford Ellis, P., *Celtic Inheritance*, Muller, London, 1985.

Berman, M., *The Reenchantment of the World*, Bantam, New York, 1984.

Blackmore, S., *Zen and the Art of Consciousness*, Oneworld Publications, London, 2011.

Bobo, G., Emmons, R.A. and McCullough, M.E., 'Gratitude in practice and the practice of gratitude' in: Linley, P.A. and Joseph, S. (Eds.), *Positive Psychology in Practice*, Wiley, Hoboken, NJ, 2004.

Bowler, P.J., *Evolution: The History of an Idea*, University of California Press, Berkeley, 1984.

Boyles, K.L., 'Saving Sacred Sites: The 1989 Proposed Amendment to the American Indian Religious Freedom Act', *Cornell Law Review*, 1991, 76, pp. 1117–48.

Braithwaite, R.B., *Scientific Explanation: A Study of the Function of Theory, Probability and Law in Science*, Cambridge University Press, 1953.

Bratman, G.N., Daily, G.C., Levy, B.J. and Gross, J.J., 'The benefits of nature experience: improved affect and cognition', *Landscape and Urban Planning*, 2015, 138, pp. 41–50.

Bratman, G.N., Hamilton, J.P., Hahn, K.S., Daily, G.C., and Gross, J.J., 'Nature experience reduced rumination and subgenual prefrontal cortex activation', *Proceedings of the National Academy of Sciences (US)*, 2015, 112, pp. 8567–72.

Brown, P., *The Cult of the Saints: Its Rise and Function in Latin Christianity*, Chicago University Press, 2015.

Brown, S., 'The "musilanguage" model of music evolution' in: N. Wallin, Merker, B. and Brown, S. (Eds.), *The Origins of Music*, MIT Press, Cambridge, MA, 2000.

Brown, S., 'Evolutionary models of music: From sexual selection to group selection' in: Tonneau, F. and Thompson, N. (Eds.), *Perspectives in Ethology*, Plenum, New York, 2004, 13, pp. 231–81.

Carter, C., *Science and the Near-Death Experience: How Consciousness Survives Death*, Inner Traditions, Rochester, VT, 2010.

Catholic Church, *Catechism of the Catholic Church*, Burns and Oates, London, 1999.

Chanda, M.L. and Levitin, D.J., 'The neurochemistry of music', *Trends in Cognitive Science*, 2013, 17, pp. 180–94.

Clift, S., Hancox, G., Morrison, I., Hess, B., Kreutz, G., and Stewart, D., 'Choral singing and psychological wellbeing: quantitative and qualitative findings from English choirs in a cross-national survey', *Journal of Applied Arts and Health*, 2010, 1, pp. 19–34.

Clift, S., Hancox, G., Staricoff, R., and Whitemore, C., 'Singing and health: a systematic mapping and review of non-clinical research',

Christ Church University Research Centre for Arts and Health, Canterbury, 2008.

Coleman, S. and Elsner, J., *Pilgrimage Past and Present: Sacred Travel and Sacred Space in the World Religions*, British Museum Press, London, 1995.

Corazza, O., *Near-Death Experiences: Exploring the Mind-Body Connection*, Routledge, London, 2008.

Cox, B., 'The Large Hadron Collider: A Scientific Creation Story' in: Sherine, A. (Ed.), *The Atheist's Guide to Christmas*, HarperCollins, London, 2009.

Cross, I., 'The nature of music and its evolution' in: Hall, M., Cross, I. and Thaut, M. (Eds.), *The Oxford Handbook of Music Psychology* (2nd edition), Oxford University Press, 2016.

Cross, I. and Morley, I., 'The evolution of music: theories, definitions and the nature of the evidence' in: Malloch, S. and Trevarthen, C. (Eds.), *Communicative Musicality*, Oxford University Press, 2008.

Darwin, C., *The Origin of Species*, Murray, London, 1859.

Darwin, C., *The Variation of Animals and Plants Under Domestication*, Murray, London, 1875.

Darwin, C., *The Origin of Species* (6th edition), Murray, London, 1882.

Darwin, C., *The Descent of Man and Selection in Relation to Sex* (2nd edition), Murray, London, 1885.

Darwin, E., *Zoonomia*, 2 vols, 1794–96, AMS Press, New York, reprinted 1974.

Davidson, L.K. and Gitlitz, D.M., *Pilgrimage: From the Ganges to Graceland: An Encyclopedia*, ABC-CLIO, Santa Barbara, CA, 2002.

Davies, J.G., *Pilgrimage Yesterday and Today: Why? Where? How?*, SCM Press, London, 1998.

Dawkins, R., *The God Delusion*, Bantam, London, 2006.

Dawkins, R., *Unweaving the Rainbow: Science, Delusion and the Appetite for Wonder*, Penguin, London, 2006.

De Botton, A., *The Pleasures and Sorrows of Work*, Hamish Hamilton, London, 2009.

De Botton, A., *Religion for Atheists: A Non-Believer's Guide to the Uses of Religion*, Hamish Hamilton, London, 2013.

Deb, D., *Sacred Groves of West Bengal: A Model of Community Forest Management*, University of East Anglia, Norwich, 2007.

Dennett, D., *Breaking the Spell: Religion as a Natural Phenomenon*, Viking, New York, 2006.

Descola, P., *Beyond Nature and Culture*, University of Chicago Press, 2013.

Ehrenreich, B., *Blood Rites: Origins and History of the Passions of War*, Metropolitan Books, New York, 1997.

Ehrenreich, B., *Dancing in the Streets: A History of Collective Joy*, Metropolitan Books, New York, 2006.

Ehrenreich, B., *Smile or Die: How Positive Thinking Fooled America and the World*, Granta Books, London, 2009.

Eire, C.M.N., *War Against the Idols: The Reformation of Worship from Erasmus to Calvin*, Cambridge University Press, 1986.

Eliade, M., *Patterns in Comparative Religion*, Sheed and Ward, London, 1958.

Emerson, R.W., *Selected Essays*, Penguin Books, Harmondsworth, 1985.

Emmons, R.A. and Crumpler, C.A., 'Gratitude as a human strength: Appraising the evidence', *Journal of Social and Clinical Psychology*, 2000, 19, pp. 56–69.

Emmons, R.A. and Kneezel, T.T., 'Giving thanks: Spiritual and religious correlates of gratitude', *Journal of Psychology and Christianity*, 2005, 24, pp. 140–8.

Emmons, R.A. and McCullough, M.E., 'Counting blessings versus burdens: An experimental investigation of gratitude and subjective wellbeing in daily life', *Journal of Personality and Social Psychology*, 2003, 84, pp. 377–89.

Evans, D., *Placebo: Mind Over Matter in Modern Medicine*, HarperCollins, London, 2010.

Feldman, R., Monakhov, M., Pratt, M., and Ebstein, R.P., 'Oxytocin pathway genes: evolutionary ancient system impacting on human affiliation, sociality, and psychopathology', *Biological Psychiatry*, 2016, 79, pp. 174–84.

Field, C.D., 'Secularising selfhood: what can polling data on the personal saliency of religion tell us about the scale and chronology of secularisation in modern Britain?', *Journal of Beliefs and Values*, 2015, 36, pp. 308–30.

Fox, M., *Meister Eckhart*, New World Library, Novato, CA, 2014.

Fox, M. and Sheldrake, R., *The Physics of Angels: Exploring the Realm Where Science and Spirit Meet*, HarperCollins, San Francisco, 1996.

Frazer, J., *Folk-Lore in the Old Testament*, Macmillan, London, 1918.

Freud, S., *Moses and Monotheism*, Hogarth Press, London, 1939.

Geissmann, T., 'Gibbon songs and human music from an evolutionary perspective', in: Wallin, N.L., Merker, B. and Brown, S. (Eds.), *The Origins of Music*, MIT Press, Cambridge, MA, 2000.

Giblett, R.J., *People and Places of Nature and Culture*, Intellect, Bristol, 2011.

Gilbert, N., 'A natural high', *Nature*, 2016, 531, pp. 556–7.

Godwin, J., *Music, Mysticism and Magic: A Sourcebook*, Routledge and Kegan Paul, London, 1986.

Goyal, M. et al., 'Meditation programs for psychological stress and wellbeing: A systematic review and meta-analysis', *JAMA Internal Medicine*, 2014, 174, pp. 357–68.

Gray, J., *Straw Dogs: Thoughts on Humans and Other Animals*, Granta Books, London, 2002.

Gray, J., *The Immortalization Commission: Science and the Strange Quest to Cheat Death*, Allen Lane, London, 2011.

Griffiths, B., *Return to the Centre*, Collins, London, 1976.

Griffiths, B., *The Marriage of East and West*, Collins, London, 1982.

H.M. Government White Paper, *The Natural Choice: Securing the Value of Nature*, The Stationery Office, London, 2011.

Hardy, A., *The Spiritual Nature of Man: A Study of Contemporary Religious Experience*, Clarendon Press, Oxford, 1979.

Harris, S., *The End of Faith: Religion, Terror, and the Future of Reason*, Norton, New York, 2005.

Harris, S., *Waking Up: Searching for Spirituality Without Religion*, Transworld, London, 2014.

Hart, D.B., *The Experience of God: Being, Consciousness, Bliss*, Yale University Press, 2013.

Hayward, G., *Singing as One: Community in Synchrony*, PhD Thesis, Cambridge University, 2014.

Hazrat Inayat Khan, 'The music of the spheres', in: Rothenberg, D. and Ulvaeus, M. (Eds.), *The Book of Music and Nature*, Wesleyan University Press, Middletown, CT, 2009.

Heller, E., *The Disinherited Mind: Essays in Modern German Literature and Thought*, Cambridge University Press, 1952.

Hitchens, C., *God Is Not Great: How Religion Poisons Everything*, Hachette, New York, 2008.

Hölzel, B.K., Ott, U., Gard, T., Hempel, H., Weygandt, M., Morgen, K. and Vaitl, D., 'Investigation of mindfulness meditation practitioners with voxel-based morphometry', *Social Cognitive and Affective Neuroscience*, 2008, 3, pp. 55–61.

Hölzel, B.K., Carmody, J., Vangel, M., Congleton, C., Yerramsetti, S.M., Gard, T. and Lazar, S.W., 'Mindfulness practice leads to increases in regional brain gray matter density', *Psychiatry Research*, 2011, 191, pp. 36–43.

Howell, P., 'At home and astray', *Cambridge Alumni Magazine*, 2016, 77, pp. 28–35.

Karjalainen, E., Sarjala, T. and Raitio, H., 'Promoting human health through forests: overview and major challenges', *Environmental Health and Preventive Medicine*, 2010, 15, pp. 1–8.

Koenig, H.G., *Medicine, Religion and Health*, Templeton Press, West Conshohocken, PA, 2008.

Koenig, H., McCullough, M.E. and Larson, D.B., *Handbook of*

Religion and Health, Oxford University Press, 2001.

Koenig, H., King, D.E. and Carson, V.B., *Handbook of Religion and Health* (2nd edition), Oxford University Press, 2012.

Kuhn, T., *The Structure of Scientific Revolutions*, University of Chicago Press, 1962.

Kuyken, W., Hayes, R., Barrett, B. et al, 'Effectiveness and cost-effectiveness of mindfulness-based cognitive therapy compared with maintenance antidepressant treatment in the prevention of depressive relapse or recurrence (PREVENT): a randomised controlled trial', *The Lancet*, 2015, 386, pp. 63–73.

La Fontaine, J.S., *Initiation: Ritual Drama and Secret Knowledge Across the World*, Penguin, Harmondsworth, 1985.

Lévi-Strauss, C., *Structural Anthropology*, Penguin, Harmondsworth, 1972.

Levitin, D., *This Is Your Brain on Music: Understanding a Human Obsession*, Atlantic Books, London, 2006.

Li, Q., 'Effect of forest bathing trips on human immune function', *Environmental Health and Preventive Medicine*, 2010, 15, pp. 9–17.

Lorimer, D., 'Science and religion: A survey of spiritual practices and benefits among European sollutions, engineers and medical professional', *Scientific and Medical Network Review*, 2017, 123, pp. 23–6

Louv, R., *Last Child in the Woods: Saving Our Children from Nature-Deficit Disorder*, Algonquin Books, Chapel Hill, 2008.

Lutz, A., Greischar, L.L., Rawlings, N.B., Ricard, M. and Davidson, R.J., 'Long-term meditators self-induce high-amplitude gamma synchrony during mental practice', *Proceedings of the National Academy of Sciences (US)*, 2004, 101, pp. 16369–72.

McCullough, M.E., Emmons, R.A. and Tsang, J.A., 'The grateful disposition: A conceptual and empirical topography', *Journal of Personality and Social Psychology*, 2002, 82, pp. 112–27.

Marais, E., *The Soul of the White Ant*, Methuen, London, 1937.

Mascaro, J. (Trans.), *The Upanishads*, Penguin Books, Harmondsworth, 1965.

Mauss, M., *The Gift: The Form and Reason for Exchange in Archaic Societies*, Norton, New York, 2000.

Merker, B., 'Synchronous chorusing and human origins', in: Wallin, N.L., Merker, B. and Brown, S. (Eds.), *The Origins of Music*, MIT Press, Cambridge, MA, 2000.

Michell, J., *The Earth Spirit: Its Ways, Shrines and Mysteries*, Thames and Hudson, London, 1975.

Mojo, *The Mojo Collection: The Ultimate Music Companion* (4th edition), Canongate Books, Edinburgh, 2003.

Morris, P.A., 'The effect of pilgrimage on anxiety, depression and religious attitude', *Psychological Medicine*, 1982, 12, pp. 291–4.

Nabokov. P., *Where the Lightning Strikes: The Lives of American Indian Sacred Places*, Penguin Books, New York, 2006.

Nagel, T., *Mind and Cosmos: Why the Materialist Neo-Darwinian Conception of Nature is Almost Certainly False*, Oxford University Press, 2012.

Nilsson, U., 'Soothing music can increase oxytocin levels during bed rest after open-heart surgery: a randomised control trial', *Journal of Clinical Nursing*, 2009, 18, pp. 2153–61.

Paffard, M., *Inglorious Wordsworths: A Study of Some Transcendental Experiences in Childhood and Adolescence*, Hodder and Stoughton, London, 1973.

Park, B.J., Tsunetsugu, Y., Kasetani, T., Kagawa, T. and Miyakazi, Y., 'The physiological effects of *Shinrin-yoku* (taking in the forest atmosphere or forest bathing): evidence from field experiments in 24 forests across Japan', *Environmental Health and Preventive Medicine*, 2010, 15, pp. 18–26.

Partridge, E., *Origins: A Short Etymological Dictionary of Modern English*, Routledge and Kegan Paul, London, 1966.

Purce, J., *The Mystic Spiral: Journey of the Soul*, Thames and Hudson, London, 1974.

Purce, J., 'Sound in mind and body', *Resurgence*, March–April 1986, pp. 26–30.

Reynolds, G., 'How walking in nature changes the brain', *New York Times*, 22 July 2015.

Rosenthal, J.Z., Grosswald, S., Ross, R. and Rosenthal, N., 'Effects of transcendental meditation in veterans of Operation Enduring Freedom and Operation Iraqi Freedom with post-traumatic stress disorder: A pilot study', *Military Medicine*, 2011, 176, pp. 626–30.

Roszak, T., *Where the Wasteland Ends: Politics and Transcendence in Post-Industrial Society*, Faber and Faber, London, 1973.

Sacks, O., *Musicophilia: Tales of Music and the Brain*, Picador, London, 2007.

Sacks, O., *Gratitude*, Picador, London, 2015.

Schwartz, S., 'Meditation – the controlled psychophysical self-regulation process that works', *Explore: The Journal of Science and Healing*, 2011, 7, pp. 348–53.

Schweiker, W., 'Torture and religious practice', *Dialog: A Journal of Theology*, 2008, 47, pp. 208–16.

Scott, C.J., Harrison, R.G., Owens, M.J., Lockwood, M. and Barnard, L., 'Evidence for solar wind modulation of lightning', *Environmental Research Letters*, 2014, 9, 055004.

Seligman, M.P., Steen, T.A., Park, N. and Peterson, C., 'Positive psychology progress: empirical validation of interventions', *American Psychologist*, 2005, 60, pp. 410–21.

Sheldrake, R., *A New Science of Life; The Hypothesis of Formative Causation*, Blond and Briggs, London, 1981.

Sheldrake, R., *The Presence of the Past: Morphic Resonance and the Habits of Nature*, Collins, London, 1988.

Sheldrake, R., *The Rebirth of Nature: The Greening of Science and God*, Century, London, 1992.

Sheldrake, R., *Dogs That Know When Their Owners Are Coming Home, And Other Unexplained Powers of Animals*, Hutchinson, London, 1999.

Sheldrake, R., *Seven Experiments That Could Change the World* (2nd edition), Park Street Press, Rochester, VT, 2002.

Sheldrake, R., *The Sense of Being Stared At, And Other Aspects of the Extended Mind*, Hutchinson, London, 2003.

Sheldrake, R., *A New Science of Life; The Hypothesis of Formative Causation* (3rd edition), Icon Books, London, 2009.

Sheldrake, R., *The Presence of the Past: Morphic Resonance and the Habits of Nature* (2nd edition), Icon Books, London, 2011.

Sheldrake, R., *The Science Delusion: Freeing the Spirit of Enquiry*, Coronet, London, 2012. (Published in the US as *Science Set Free: Ten Paths to New Discovery*, Random House, New York.)

Sheldrake, R. and Fox, M., *Natural Grace: Dialogues on Science and Spirituality*, Bloomsbury, London, 1996.

Sheldrake, R., McKenna, T., and Abraham, R., *The Evolutionary Mind: Trialogues at the Edge of the Unthinkable*, Trialogue Press, Santa Cruz, CA, 1998. (New edition, Monkfish Books, Rhinebeck, NY, 2005.)

Sheldrake, R., McKenna, T., and Abraham, R., *Chaos, Creativity, and Cosmic Consciousness*, Park Street Press, Rochester, VT, 2001.

Sheldrake, R. and Smart, P., 'Videotaped experiments on telephone telepathy', *Journal of Parapsychology*, 2003, 67, pp. 147–166.

Smith, J.K.A., *How (Not) To Be Secular: Reading Charles Taylor*, Eerdmans, Grand Rapids, MI, 2014.

Spencer, N., *Atheists: The Origin of the Species*, Bloomsbury, London, 2014.

Steindl-Rast, D., *Gratefulness, The Heart of Prayer*, Paulist Press, Mahwah, NJ, 1984.

Strawson, G., 'Realistic monism: why physicalism entails panpsychism', *Journal of Consciousness Studies*, 2006, 13, pp. 3–31.

Taylor, C., *A Secular Age*, Harvard University Press, Cambridge, MA, 2007.

Taylor, E., Introduction in: Murphy, M., Donovan, S. and Taylor, E., *The Physical and Psychological Effects of Meditation: A Review of Contemporary Research with a Comprehensive Bibliography, 1931–1996*, Institute of Noetic Sciences, Sausalito, CA, 1997.

Bibliography

Theos, *The Spirit of Things Unseen: Belief in Post-Religious Britain*, Theos, London, 2013.

Thomas, K., *Man and the Natural World: Changing Attitudes in England 1500–1800*, Penguin Books, Harmondsworth, 1984.

Titze, I.R., 'The human instrument', *Scientific American*, 2008, 298 (1), pp. 94–101.

Tolkien, J.R.R., *The Silmarillion*. HarperCollins, London, 2013.

Tomlinson, G., *A Million Years of Music: The Emergence of Human Modernity*, Zone Books, New York, 2015.

van Gennep, A., *The Rites of Passage*, Chicago University Press, 1972.

van Lommel, P., *Consciousness Beyond Life: The Science of the Near-Death Experience*, HarperOne, London, 2011.

Vella-Burrows, T., 'Singing and people with dementia', in: Clift, S. (Ed.), *Singing, Wellbeing and Health*, Christ Church University Research Centre for Arts and Health, Canterbury, 2012.

Vitebsky, P., *Reindeer People: Living with Animals and Spirits in Siberia*, Harper Perennial, London, 2011.

Viveiros de Castro, E.B., 'Exchanging perspectives: The transformation of objects into subjects in Amerindian ontologies', *Common Knowledge*, 2004, 10, pp. 463–84.

Watkins, P.C., van Gelder, M., and Frias, A., 'Furthering the science of gratitude', in: Lopez, S.J. and Snyder, C.R. (Eds.), *The Oxford Handbook of Positive Psychology* (second edition), Oxford University Press, New York, 2009.

Wax, R., *A Mindfulness Guide for the Frazzled*, Penguin, London, 2016.

Williams, D.C., *God's Wilds: John Muir's Vision of Nature*, Texas A&M University Press, 2002.

Wilson, E.O., *Biophilia*, Harvard University Press, 1984.

Woodard, G.D. and McCrone, W.C., 'Unusual crystallization behavior', *Journal of Applied Crystallography*, 1975, 8, p. 342.

Woodhead, L. and Catto, R., *Religion and Change in Modern Britain*, Routledge, London, 2012.

Wroe, A., *Being Shelley: The Poet's Search for Himself*, Vintage, London, 2007.

Young, J., McGown, E. and Haas, E., *Coyote's Guide to Connecting with Nature*, Owlink Media, Shelton, WA, 2010.

Zivotofsky, A.Z., Gruendlinger, L., and Hausdorff, J.M., 'Modality-specific communication enabling gait synchronization during over-ground side-by-side walking', *Human Movement Science*, 2012, 31, pp. 1268–85.

Zuckerman, P., 'Atheism: Contemporary numbers and patterns', in: Martin, M. (Ed.), *The Cambridge Companion to Atheism*, Cambridge University Press, 2007, pp. 47–65.

Index